Leading Change
IN A
Web 2.1 World

INNOVATIONS IN LEADERSHIP

The Innovations in Leadership series, a collaboration of the Olin Business School at Washington University in St. Louis and the Brookings Institution Press, offers books that are succinct, action-oriented, and pragmatic, focused on a wide range of problems facing business and government leaders today. Each title will offer practical approaches and innovative solutions to meet present-day challenges.

Washington University and the Brookings Institution share the indelible stamp of philanthropist and businessman Robert S. Brookings, an innovator in responsible private enterprise and effective governance, who founded the Brookings Institution and was a major benefactor of the university and chairman of its board of trustees. The series seeks to emulate Brookings's legacy to seek and inform in order to achieve a balance in the public and private sectors leading to improved governance at all levels of American public and private life.

Leading Change
IN A
Web 2.1 World

How ChangeCasting Builds Trust,
Creates Understanding, and
Accelerates Organizational Change

Jackson Nickerson

BROOKINGS INSTITUTION PRESS
Washington, D.C.

Copyright © 2010
THE BROOKINGS INSTITUTION
1775 Massachusetts Avenue, N.W., Washington, D.C. 20036
www.brookings.edu

First paperback edition 2013

The Library of Congress has cataloged the hardcover edition as follows:
Nickerson, Jackson A.
  Leading change in a Web 2.1 world : how ChangeCasting builds trust, creates understanding, and accelerates organizational change / Jackson Nickerson.
    p.   cm.
  Includes bibliographical references and index.
  Summary: "Provides a new personnel management model called ChangeCasting, based on Web 2.0 technology, that organizations can use to bring about change and encourage more dynamic organizational structure and communications, with examples of how successful the model has been in practice"—Provided by publisher.
  ISBN 978-0-8157-0484-3 (hbk. : alk. paper)
  1. Organizational change. 2. Organizational behavior. 3. Web 2.0. 4. Information technology—Management. 5. Management. I. Title.
HD58.8.N497 2010
658.4'06—dc22   2010030573
          ISBN 978-0-8157-2542-8 (pbk. : alk. paper)

Digital printing

Printed on acid-free paper

Typeset in Sabon and Ocean

Composition by Oakland Street Publishing
Arlington, Virginia

CONTENTS

To my parents
who taught me the importance of good communication

PREFACE

How many times have you attempted to change your organization—whether the entity is a team, division, company, government agency, or nonprofit? Like most organizations, yours probably faced severe downsizing in response to the financial crisis of 2008. Reducing personnel, retracting from markets, and scaling down your organization must have presented a difficult leadership challenge. Perhaps in better times you were assigned the task of implementing a new information system, manufacturing procedure, or customer-service process. Or perhaps you had identified a problem or opportunity within your organization and wanted to respond to it by driving costs down or grabbing an opportunity to grow revenue. Or your change efforts may have been on a larger canvas, such as fundamentally repositioning your organization to survive the downturn, capture competitive advantage, locate new resources, or respond to the head office's decision to restructure your division. Whatever attempts you made to bring about change and paths you traveled to accomplish your objectives, you probably wished your change efforts had progressed faster, been easier, and produced better results. If you are like most leaders, you probably wish that your community had more eagerly embraced the change, responded to the challenge, and seized the opportunity.

This book argues that recent advances in Web 2.0 technology enable new leadership processes and guidelines that can create value for organizations. The term "Web 2.0 technology" refers to

> Web applications that facilitate interactive information sharing, interoperability, user-centered design, and collaboration on the World Wide Web. A Web 2.0 site allows its users to interact with each other as contributors to the website's content, in contrast to websites where users are limited to the passive viewing of information that is provided to them. Examples of Web 2.0 include web-based communities, hosted services, web applications, social-networking sites, video-sharing sites, wikis, blogs, mashups, and folksonomies.[1]

So what, then, is Web 2.1? I refer to the combination of processes and guidelines that utilize Web 2.0 technology that change people's behavior as Web 2.1. The aim of using the specific processes and guidelines described in this book is not just leading change but accelerating it. I coined the term "ChangeCasting" to refer to the combination of Web 2.0 technology and the specific set of processes and guidelines described in this book.

Generally speaking, Web 2.0 technology facilitates creativity, collaboration, and sharing among users of digital devices such as computers and smart phones. ChangeCasting is a process and set of guidelines using these media to accomplish something else—to enable you to build trust and create understanding in novel ways, which, when combined with a variety of leadership approaches, can help you accelerate organizational change. Although the emergence of Web 2.0 technology is a precondition for the ideas presented in *Leading Change in a Web 2.1 World*, this book is not about technology. It is about helping you become a better leader of people. People are the roots of your organization's success—but roots often are difficult to dig up, move, and replant. Hence, organizational change can be difficult, costly, and risky.

Leading Change in a Web 2.1 World provides some new insights into why it is so difficult to engage people and organizations in change. It explains how web-based video communications, when used in accordance with ChangeCasting's specific process and guidelines, is a key way to build trust and create understanding, thereby unlocking and accelerating organizational change.

The idea of using video communications within an organization is not new, but the process of ChangeCasting is new because it explains how to use information technology to lead change. In the pages that follow, you will read the stories of two real CEOs who faced tremendous organizational challenges in reconfiguring their firms to pursue new business strategies. One CEO adopted the use of web-based videos but not ChangeCasting's process and guidelines. The other adopted the use of web-based videos PLUS the ChangeCasting process and guidelines. The latter achieved remarkable success in accelerating organizational change, while the former's efforts geared to change became mired in organizational resistance; even after several years of implementation efforts, successful change remained elusive.

A book about using Web 2.0 technologies for leading change would be incomplete without some reflections on the 2008 presidential election. Many factors came into play in Barack Obama's victory over John McCain, but among them, the use of Web 2.0 technologies was striking. I purchased an iPhone in September 2008, and later downloaded the Obama campaign "app" (application) to see how it worked and what features it offered. I was stunned to find, while walking home a couple of days before the election, that with the push of a digital button, I was able to watch on my iPhone, up close and personal, a two-minute video clip of Obama speaking in my home state of Missouri just hours before. Watching this video on a digital phone was surprisingly compelling. The McCain campaign offered no equivalent communication approach.

Obama's use of this technology did not stop with the campaign. In fact, one could reasonably assert that he essentially used "ChangeCasts" during the presidential transition by providing weekly videos for citizens to watch and respond to. These weekly videos typically were about two to three minutes in length. Even after occupying the White House, the president continued to provide ChangeCasts in his attempt to lead change for an entire nation. Although the Obama campaign didn't invent the idea of ChangeCasting, it certainly applied the use of Web 2.0 technologies to leading change and brought the application of these technologies to a new level—a 2.1 level. The use of Web 2.0 technologies is here to stay. The question for you is how to best take advantage of these technologies to lead change in your community.

This book provides one answer. Many people are nervous about being filmed or appearing in a video, especially in a work context. ChangeCasting is designed to help you overcome such fears by providing a specific process and guidelines to eliminate the uncertainty that stokes fear. The particular process and guidelines of ChangeCasting guide you through the who, what, when, where, and how of web-based video so that you can use it to build trust and create understanding in order to accelerate organizational change. By following the process and guidelines described in this book you can enhance your success as a leader, which can have an important impact on your community as well as your career. I hope that you enjoy reading this book and that it creates value for you and your community.

Although every book has an author's name printed on the cover, many people contribute to its creation. I am indebted to the numerous people who have contributed to the writing of *Leading Change in a Web 2.1 World*. First and foremost I thank the two CEOs (whose real names are not used in this book) who adopted Change-Casting in some of its first settings and whose cases I use in the following chapters as examples. I learned much about the practical application of ChangeCasting processes from them and am deeply indebted to them for their participation.

Early drafts of the first few chapters were read by many students in the MBA and Executive MBA programs of the Olin Business School at Washington University in St. Louis. Their reactions helped to refine concepts and language used to convey the process and guidelines. I also thank participants of Olin's Thought Leadership conference, who provided early reactions to ChangeCasting and offered thoughts on how they could use it in their organizations. Many faculty members at the Olin Business School—including Stuart Bunderson, Kurt Dirks, Ray Sparrowe, and Todd Zenger—not only provided useful reactions to the ideas presented here but pushed me to improve my understanding of organizational change. Brian Silverman, from the University of Toronto, also provided useful feedback. Dick Mahoney pushed my thinking on many aspects of the text and suggested the book's title. Without such collegial interactions and encouragement, *Leading Change in a Web 2.1 World* might not have been written.

Tim Gray is a friend who provided wonderful and valuable insights into the publishing world. Brad Fels is a friend and business partner who was an early collaborator on this project. I learned much from him about video communication and being creative.

I owe many thanks to Brookings Institution Press and its publication team of Christopher Kelaher, Janet Walker, and Susan

Woollen, and especially Katherine Scott, whose editorial help and assistance greatly improved the content of this book. Rich Pottern designed a very fitting and exciting graphic for the book cover. I especially appreciate the support of Robert Faherty, director of Brookings Institution Press, who was willing to take the risk of launching a new book series, Innovations in Leadership, with *Leading Change in a Web 2.1 World* as its first installment.

This book, along with much of my professional life, would not have been possible and not nearly so rewarding if not for my family. Cici, Genevieve, and Will bring light and joy into my life. They have suffered with great patience through several years as I wrote *Leading Change in a Web 2.1 World*. I hope they believe that their patience was worth it.

1

Introduction: A New Tool for Leaders

After more than thirty years with the same company, Genevieve (Gen) Laneau has gotten her chance to captain the ship. And her ship is neither small nor easy to sail. Production is global, distribution is worldwide, and the Internet, among many distribution channels, plays an important and growing role for both her firm and her competitors. A multibillion-dollar technology enterprise with more than seven thousand workers spread across thirty-five countries, her company is a typical midsize global firm.

Several competitive storms were on the horizon when Gen took over as CEO. Recently consolidated competitors were expanding, growing revenues, and encroaching on the unique market position of her firm, WorldCo. Indeed, WorldCo had once been known as the high-tech leader in its field, but now, customers were turning to Gen's competitors for new products, services, and solutions. The previous CEO had successfully increased margins by focusing on operational excellence, but it was clear to Gen that focusing on cost reductions alone was not going to keep her firm ahead of the competition. She was not alone in her conclusions about the firm's future. Wall Street could see the competitive landscape and had

1

come to the same conclusions. Without new sources of profitable revenue growth the firm's stock price was not going to appreciate. The handwriting was not just on the wall, it was in the analysts' reports.

Gen was facing the leadership challenge of her career. She could see that if she didn't take action, the day of reckoning in the form of stiff competition and low profit margins was fast approaching. She realized she had to change her organization all the way to its core if her company were to overcome stagnant revenue growth and simultaneously maintain its industry-leading return on equity.

William Tracey faced a different challenge. He had joined his manufacturing firm, MandACo, more than fifteen years before being appointed CEO a few years ago, having risen largely through its financial ranks. He shepherded a Fortune 500 firm with more than two hundred plants and thirty thousand employees. William's firm became the industry's largest player through a consolidation strategy that was largely funded by taking on more debt. In the growing economy of the 1980s and 1990s, the firm's mergers and acquisitions grew revenues and profits. Although the debt load was great, economies of scale in purchasing led to profits that far exceeded interest payments.

By the twenty-first century, two trends had taken hold that changed the fundamental environment in which MandACo competed. First, customers began moving production operations to China, where MandACo did not have substantial operations. This shifting of business across the Pacific decreased demand in North America. This shift in demand had the greatest impact on MandACo, the largest manufacturer in the industry. Second, North American customers either remained small "local" customers or had become large "national" accounts. These national accounts created commodity markets by auctioning off their demand and playing suppliers off against each other. Some of MandACo's competitors had chosen long ago to focus on these national accounts and had become low-cost producers. Smaller competitors focused on local customers and differentiated themselves from their competition because of their speed, flexibility, and customized service. MandACo's plants continued to sell to both types of customers, thereby catering specifically to neither type, which put his firm at a competitive disadvantage.

By the time William became CEO of MandACo, the company was unprofitable and its financial condition was actually worsening, resulting in a drop in the stock price of 30 percent from its peak. His first response was to shut down manufacturing facilities and take capacity out of the market, but demand shrank faster than he could close capacity. With little financial room to maneuver and still needing to make debt payments, he was unable to invest in and update production equipment. His multibillion-dollar firm was unprofitable and its financial condition was getting worse. As some observers wondered whether his highly leveraged firm would survive, MandACo's survival—and William's career—depended on his ability to fundamentally change and revive his organization.

The Test of Leadership

Managing organizational change is not a challenge for just Gen and William. Organizational change is a common challenge; moreover it is *the* test of leadership, because failure is so often the outcome. A 1998 study by Wheatley and Kellner-Rogers reported on a survey of chief executive officers who stated that up to 75 percent of organizational change efforts do not yield the promised results. A 2002 study undertaken by Miller estimated that change initiatives critical to organizational success typically fail 70 percent of the time. In a 2004 study, Raps found a 70 to 90 percent failure rate in companies' attempts to implement new strategic plans.[2] These studies and others leave no doubt that leading organizational change is difficult, costly, and risky. Failure is the norm, but the benefits of successful organizational change are great. For some companies it is change or die. For others, successful organizational change can rejuvenate organizations and catapult CEOs into the pantheon of great business leaders who revived their companies' growth and profitability and earned themselves fame and fortune in the process.

Some business leaders already in this pantheon are Allied Signal's Larry Bossidy, Monsanto's Dick Mahoney, GE's Jack Welch, IBM's Lou Gerstner, and Emerson's Chuck Knight. Each of them took a large company that was performing poorly or even losing money and rebuilt it to achieve great market and financial success. Larry Bossidy is well known for taking Allied Signal, a collection of not very successful businesses that was losing money, and turning them around until they were acquired by Honeywell in 1999. Dick Mahoney rebuilt the chemical company Monsanto and repositioned it in the life sciences with investments in agricultural biotechnology and the acquisition of G. D. Searle Pharmaceuticals. The story of General Electric's Jack Welch, who reconstructed a financially hemorrhaging giant and turned it into a global juggernaut, is well known (Larry Bossidy worked for Jack Welch at one time). Lou Gerstner is credited with saving the global paragon IBM from going out of business in the early 1990s. Chuck Knight of Emerson Electric took a small manufacturer and delivered one of the longest profitable growth trajectories in the history of Wall Street.

Nowadays, leading organizational change is more difficult than ever before. The ongoing revolution in information technology is increasing the frequency of organizational change and making it harder. Thomas Friedman noted in *The World Is Flat: A Brief*

History of the Twenty-First Century (2005), that the revolution in information technology, among other factors, has led to a new level of globalization and integration. The Internet has indeed connected the world. Now practically any company can quickly design products in the United States, manufacture them in China, provide customer service from India and the Philippines, and serve customers on every continent. But so can its competitors—not just its old local competitors but also companies in almost any region of the globe. Information technology and myriad outsourcing organizations bring down entry costs, which invites increased competition. Markets are becoming increasingly commoditized, meaning that it is difficult for firms to differentiate themselves from one another. Firms that once had profitable business strategies now face cut-throat global competition.

Many leaders are responding to these challenges by hammering flat their organizational structures. They are dispersing their operations and delegating and decentralizing decisionmaking while at the same time spreading their firms' activities around the world and physically farther apart as they seek to drive down costs and take advantage of local capacities. Leaders are reconfiguring their organizations—acquiring organizations, selling off portions of their companies—to find new sources of advantage. Yet these survival tactics are also making the test of leadership more demanding. These widely dispersed, flat, and reconfigured organizations may be better able to adapt quickly to local conditions, but they are far more difficult to coordinate, integrate, manage, and lead than vertically structured command-and-control-style organizations.

These pressures require organizations, and their leaders, to respond and adapt to markets as never before. Business strategies used to be effective for at least a decade, if not longer. Now, changes in business strategy every two or three years are not unusual. Any misstep or delay in adapting, reconfiguring, and responding to customers' needs allows competitors quickly to drain away customers, revenue, and profit. Staying ahead of the competition requires frequent organizational change. The old saw that organizational change is the only constant takes on new meaning in the context of twenty-first-century global competition.

Nonprofits and government agencies face similar pressures. Financial pressures from shrinking budgets and increasing demands from funders and constituents all are magnified by the Internet because it makes it easier and faster for the demands to manifest themselves, creating a driving impetus for change. Few organizations are insulated from these pressures.

How can you accelerate change in your organization to stay ahead of the competition and respond to these pressures? How can you effectively lead organizational change in vertically structured organizations or in organizations whose workers are geographically dispersed, encouraged to act locally, and told to take responsibility and fight bureaucracy? Whether you are planning a minor initiative or a major strategic overhaul, how are you going to capture your workers' minds and hearts so that they are motivated to embrace change and alter their behavior? Leading organizational change—the test of leadership—has never posed a greater challenge.

Leading Change in a Web 2.1 World describes a new approach for accelerating organizational change that leverages Web 2.0 technology by combining it with particular leadership processes and guidelines to create a Web 2.1 methodology. This method is called ChangeCasting. To reiterate: ChangeCasting is a process and a set of guidelines for communicating with and listening to your community, for creating a conversation that can lead to vital changes in your community's behavior. ChangeCasting uses the same revolutionary technologies that lie at the root of the problem, information technology and the Internet, to provide a new tool

FIGURE 1. The ChangeCasting Process

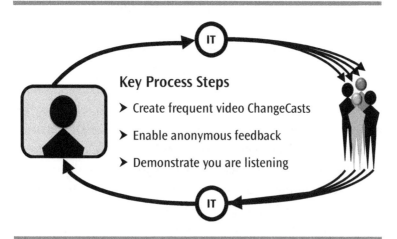

Key Process Steps

➤ Create frequent video ChangeCasts

➤ Enable anonymous feedback

➤ Demonstrate you are listening

that can help you accelerate change. Becoming a ChangeCasting leader—by adopting the specific process and guidelines described in this book—enables you to launch a conversation with your entire community that can build trust, create understanding, and accelerate organizational change.

How ChangeCasting Works

The process of ChangeCasting is built around the use of web-enabled and webcast videos, called "ChangeCasts," launched from the Internet or from an organization's Intranet (see figure 1). An Intranet is a restricted computer network, a private network created using World Wide Web software. Using Web 2.0 information technology (IT), the leader creates a video that is available for viewing by the leader's entire community, wherever members are located, whenever they choose to view it.

But merely sending a video is not enough to stimulate a conversation. For a good conversation to take place, leaders must listen, hear their community's response, and demonstrate they have listened—in short, must engage in a two-way dialog. Leaders must explicitly invite their community to provide honest reactions and feedback and pose questions about the message in such a way that the community does not fear reprisals. Here again, Web 2.0 information technology can help. A leader can stimulate a two-way conversation by inviting community members to send their reactions to ChangeCasts anonymously. The process of inviting feedback through anonymous responses is necessary to the process; without anonymity, fear will prevent expressions of distrust and misunderstanding. The vital components of the ChangeCast process are leaders' sending messages through ChangeCasts, listening to the community's anonymous responses, and demonstrating, by means of the next ChangeCast, that they have listened.

When leaders receive feedback, they must acknowledge that they have listened and understood what they have heard. Using web-enabled video allows leaders to rely on a full spectrum of communication techniques to display their level of understanding. Perhaps more important, video enables the leader to demonstrate understanding in ways not readily available by e-mails, blogs, memorandums, and the chain of command. For instance, leaders can visually show the community the challenges the company faces. They can prepare videos that feature workers and groups supporting the organizational change. They can show, not just tell, how the change efforts are having an impact. And the community can see facial expressions and body language that can communicate what words and audio alone can't.

In addition to the three key process steps, ChangeCasting uses specific guidelines relating to the creation of the message, the delivery of the message, and the use of video:

These guidelines help you choose messages, deliver them, and create images to build trust and increase understanding. When it

ChangeCasting Guidelines

The Message	The Delivery	The Video
Be brief and ChangeCast on a regular schedule.	Be yourself.	Speak to the camera eye-to-eye.
Use one idea for each ChangeCast.	Be compassionate.	Be close to the camera.
Be real about today.	Be direct.	Be distant from the background.
Aspire for tomorrow.	Invite conversation.	Dress for digital images.
Formulate the problem before trying to solve it.	Use symbolism.	Choose the right audio and light.

comes to choosing messages, the ChangeCasting approach stresses that to be effective, videos must arrive periodically, on a regular, ongoing basis every few weeks. Present just one main idea in each ChangeCast, and be frank and realistic about the challenges of today while offering aspirations for what your community and organization should look like or how it should operate in the future. Messages should not progress from a discussion of the community's challenge to framing the challenge *until* understanding of the issue is widespread. Only much later in the process should you move on to offering solutions and attempt to implement them.

In delivering these messages it is important for the leader to be authentic, passionate, and direct. The ChangeCasting approach incorporates ways to invite conversation with the community—in particular, the benefits of visually showing symbols of investment of capital and human assets in the organizational change effort. This signals commitment to the community and gains members' commitment.

The correct video guidelines are also a component of Change-Casting: how to speak to the camera and choose clothes so that you look good in digital images. In the directorial role behind the camera, ChangeCasting also prescribes appropriate locations and framing of your communications to help you fully engage your community. In this book I also review alternative technologies for delivering ChangeCasts and receiving anonymous feedback. The aim is to teach you to use information technology with ease and to keep it affordable.

Why ChangeCasting?

Over the past decade or so, people have become familiar with and use the ongoing innovations in electronic communications especially with respect to video. Video communication is not new to most communities. Indeed, people now expect to receive important information through web-enabled video and often give it more attention than they give to e-mails and memorandums. Many people now receive their news through web-enabled videos from CNN, the *New York Times*, MSN, Bloomberg News, and other web-enabled news sources. Millions of people use video-enabled Internet sites such as YouTube to share and gather information. Employees receive training through web-based video. Skype, Office Communicator, Adobe Connect, ooVoo, and other software that provides video and audio communications are beginning to replace telephones. In numerous contexts, society is rapidly shifting to web-enabled video as a primary means of receiving and processing information. So what is new about ChangeCasting? Why compete with professional news anchors?

Video is a medium that has become widely available and used for transmitting information; the use of video is not unfamiliar to leaders. But ChangeCasting offers leaders a novel way to use video; ChangeCasting's processes and guidelines are new. Using them, leaders can engage in two-way conversations to build trust and create understanding, which are necessary steps for accelerating organizational change. Other means of communication alone cannot reliably replace fear with trust, and uncertainty with understanding. ChangeCasting enables you to accelerate organizational change, whether your organization is vertically structured or widely dispersed and flat; whether it is small, medium, or large. ChangeCasting helps you lead your organization to accomplish your goals, stay ahead of the competition, and operate at peak performance.

This process and these guidelines allow you to quickly spot community concerns and misunderstandings; these can then be dealt with in the next video, thus stimulating an ongoing conversation. By initiating an ongoing exchange, you quickly hear about and can respond to rumors that, very often, work against change efforts. Thus, you can prevent the development of an information vacuum, which if left alone will typically fill with false, but often believable, information.

One of the great advantages of ChangeCasting is that it can be used to reach all members of a far-flung community quickly, frequently, and inexpensively, whether they are dispersed across a facility, the country, or the globe. Community members can view messages without interrupting work and can replay and review them at their convenience, to reinforce their understanding and formulate reactions and questions.

A further plus with this approach is that the community is encouraged to communicate back to you without inhibition, without the fear that suppresses honest exchanges in meetings and groups. You can go beyond inviting comments and reactions, to ask for examples of successful implementation of organizational change. This gives you the opportunity to include these examples in future ChangeCasts, thus making heroes of those who adopt the change and encouraging others to follow suit. In this way, the

ChangeCasting process creates conversations that accelerate change across a leader's entire community.

What ChangeCasting Is Not

ChangeCasting is not about being a star; it is not about developing charismatic or inspirational leaders. It is not about developing personality traits to change leadership styles. Nor does it offer a way to persuade a community that the leader's logic is right, to get buy-in on management's goals, or to generate a temporary feel-good solution. ChangeCasting is not a panacea for dealing with these and other leadership issues. It is a tool for leading change by reaching the minds and hearts of your community and for encouraging behavioral change by reducing fear and uncertainty. ChangeCasting is not intended to be used in isolation or to replace other change management processes. In fact, it is designed to complement and accelerate just about any change management process.[3]

One of the two CEOs introduced earlier became a ChangeCasting leader shortly after announcing a major organizational change (whether it was Gen or William is revealed later in the book). This leader's ChangeCasts, each one two to five minutes long, were released every two weeks. The leader used the messages to launch a conversation about the firm's new direction. Community members were able to see, via web-enabled video, this leader's emotion, energy, and commitment to the firm's new direction. The leader also showed where and how investments were being made to grow the firm in ways consistent with the new strategy. Employees who embraced the change were made heroes in later videos, which further encouraged feedback and participation. Every ChangeCast discussed some of the high volume of feedback and questions received each week from employee e-mails and surveys. In fact, every computer at the firm now has an icon on its desktop that when pressed launches an e-mail to the CEO to make it easy for everyone, from janitors to vice presidents, to ask questions, offer reactions, and engage in the conversation.

This CEO used ChangeCasting to win over the minds and hearts of the community by building trust and creating understanding about the direction of the firm, and in response the community accelerated organizational change, which in turn led to positive changes in the company's financial profile (quantitative results are presented in chapter 10).

The other CEO also tried using webcasts, but this leader did not adhere to ChangeCasting's process and guidelines. The leader's webcasts were long, stagy, and ineffective. No anonymous feedback mechanism was made available, so communications occurred in only one direction. Even four months after the reorganization was announced, many midlevel managers remained unclear about the new strategy, the specifics of the organizational change, and what they should be doing to help the company become successful. During this time the organization remained uncertain and fearful. There is no doubt that the CEO worked hard to communicate the message of change. Yet, these communications did not fully and effectively penetrate the community to generate trust, understanding, and commitment to changes in behavior. The CEO did not fully capture the minds and hearts of the community, and this was, ultimately, reflected in the financial results (see chapter 10).

Choosing to become a ChangeCasting leader does entail risks. No approach can be guaranteed to succeed, for it may be used incompletely or improperly. A video message that does not implement the ChangeCasting processes and guidelines can generate mistrust and amplify misunderstanding. Used imprudently, ChangeCasting, like any other change management tool, can fail to accelerate change and could slow it. The purpose of *Leading Change in a Web 2.1 World* is to explain to you just how to use the approach to help you advance your leadership skills, on the premise that doing so will earn returns for your organization and for you. The book describes the benefits of ChangeCasts and details the specific processes and guidelines you need to become a successful ChangeCasting leader who can accelerate organizational change and pass the test of leadership with flying colors. My sincere hope is that this book will enhance your ability to lead change through ChangeCasting and will make your leadership more enjoyable, more effective, and more worthwhile.

Chapter 2 describes the situation facing Gen and William, two CEOs leading organizational change to reinvent their companies. Gen's company already is successful by almost every financial performance measure, but she sees storm clouds on the horizon and adopts a new strategy in anticipation of the changing environment. Given its past success, Gen needs to overcome those who are resistant to change and who don't understand why the organization must change. William's company is financially distressed—unprofitable and carrying a heavy debt burden. Employees know change must occur and are waiting for it. William adopts a new strategy and proceeds to implement it. Although the two companies' situations differ, Gen and William face the same problem—a problem that bedevils all managers, not just CEOs: How can they get their communities to move together in the same direction so that their companies can change quickly and maintain or achieve high performance?

Chapter 3 explores the reasons why organizational change is so difficult and spotlights the most important barriers to organizational change. Leaders communicate all the time through e-mails, memos, the chain of command, meetings, public forums and speeches, corporate newsletters, and conference calls to analysts.

Why do these messages often fail to generate understanding and change community behavior? Why do community members distrust and misunderstand the messages and fail to incorporate the message into every decision and action they make on behalf of the organization?

Chapter 4 presents the necessary steps a leader must make to enable change. In particular, it offers two alternative approaches to enabling change: one that is based on compliance and the other based on commitment. Each approach is described in detail along with the costs and benefits of its adoption. ChangeCasting relies on the commitment approach. The chapter introduces three principles that, if put into practice, enable organizational change by creating trust and developing understanding to overcome fear and uncertainty:

1. Converse with the members of your community all at once.
2. Protect your community's conversation.
3. Show that you listen to your community.

These principles provide the foundation for ChangeCasting's effectiveness.

Chapter 5 introduces the three primary ChangeCasting process steps—corresponding to one of the three practices just enumerated—and describes how they both enable and accelerate organizational change. Thus, ChangeCasting enables a leader to communicate with all of his or her community members in a way that simply wasn't feasible or was too costly before the advent of modern information technology systems. Providing web-enabled anonymous feedback supports protecting the conversation. You show that you listen to your community by responding, in your next ChangeCast, to vital questions raised by feedback. The chapter provides specific guidance on carrying out these process steps so that a leader can readily implement them. The chapter is a handy checklist, to make the adoption of ChangeCasting easier.

Chapter 6 introduces the first of three sets of ChangeCasting guidelines: how to shape messages to engage in a conversation with your community. It explores the length and content of messages and how messages should change over the life of the change effort. Message length is central: most leaders speak far too long. Evidence is provided from several online communication sources that messages should be around two minutes long and should not exceed five minutes. Brevity is more easily achieved when each message has a single theme.

Leaders should be realistic about the current situation, yet conclude each message with their aspiration for tomorrow. Another important guideline is to use many early messages to converse about the community's challenge and to formulate it. Only after the community understands and agrees on the description of the challenge should the leader begin to seek out solutions and initiate implementation. This framework helps leaders think about what messages will help accelerate change and which messages will inhibit it.

Chapter 7 is about how to deliver messages. No one wants to be embarrassed when beaming their image and message to their entire community. In fact, fear may stop many from becoming a Change-Casting leader. This chapter presents five guidelines to help you improve the delivery of your communication. For instance, leaders are encouraged to be authentic, compassionate, and direct in their ChangeCasts. Chapter 7 provides tips and tricks on how to encourage feedback and conversation and highlights ways leaders can use symbolism to build trust and develop understanding and commitment from their communities.

In chapter 8 I discuss video, including basic guidelines for using a camera or webcam to frame the leader to make sure that he or she looks good on video; speaking to the camera; dressing to look good in digital images; and the leader's distance from both the camera and the background.

Chapter 9 provides a brief overview of the information technology (IT) available for ChangeCasting, reviewing video-capture devices and both simple and advanced Web 2.0 technologies that you may want to consider adopting. The chapter summarizes the available technologies and relative costs of these technologies—the advantages and disadvantages of low-, moderate-, and high-cost alternatives for implementing ChangeCasts. Technologies for all levels of IT expertise and all budgets are available for you to begin ChangeCasting.

Chapter 10 returns to the case studies of Gen and William. One of them became a ChangeCasting leader and one did not. The chapter describes how each communicated to her or his community, how the community responded, and what happened to each company. It draws on these illustrations to describe how ChangeCasting can be a vital—even a central—element in accelerating organizational change. The chapter also describes how ChangeCasting is not just for CEOs. You can become a ChangeCasting leader no matter what position you hold in your community and no matter the size of your community: whether you are a supervisor for a group of nurses or the president of the United States of America. ChangeCasting offers a new approach to holding conversations with your community. The chapter also spotlights several environments in which ChangeCasting can be particularly valuable.

Chapter 11 offers several "next steps" for you to take for leading change in a Web 2.1 world. Everything you need to know to become a ChangeCasting leader is contained in this book, but you may want additional resources, and these are available on the ChangeCasting webpage (www.ChangeCasting.com). Resources include actual ChangeCast dos and don'ts and information on Web 2.0 hardware and software alternatives. The website also has a link to a service where you can send a sample of your ChangeCasts to receive feedback on how you are doing.

May you find the rest of *Leading Change in a Web 2.1 World* enjoyable to read and that it helps you advance your skills for changing and leading your community. My hope is that *Leading Change in a Web 2.1 World* can help you excel in your fundamental test of leadership.

2

Managing Change: The Fundamental Test of Leadership

Both Gen and William have recently been appointed CEOs of their respective companies, as was described at the beginning of the previous chapter. Gen had been with her company for thirty years, William for just fifteen. The challenges they faced were diverse, and both newly minted CEOs decided they had to lead their companies through difficult transitions and transformations. For both, leading change would be the biggest challenge they had faced in their careers.

Gen's Challenge

More than a year earlier the board of directors selected Gen over several other candidates to fill the shoes of the then CEO, George, when he decided to step down.

George had been a successful CEO. Under his stewardship the company had focused on cost reductions and had done so very successfully. Its stock price had doubled in five years, even through the dot-com bust. Return on equity (ROE) had reached the targeted level of 20 percent. The company had relatively low debt and a strong balance sheet. With these financial results the company stood head and shoulders above the rest of its industry.

To consistently generate these results George focused not only on controlling costs but also, more important, on driving them down. George was known for penny pinching—even the hiring of a new janitor at a distant production plant could not proceed without his approval. Thus, the centerpiece of George's profit strategy was incessant cost reduction through operational excellence, not just controlling cost creep.

Using a process for identifying opportunities, he ferreted out every imaginable cost reduction—and lower costs he did. Over the six years of his tenure George lowered the operating costs and cost of goods sold by more than 10 percentage points. In fact, almost all of the five-year doubling of stock price could be attributed to his cost reductions as revenue growth unfortunately remained in the low single digits.

The financial success of George's strategy was undeniable—it was the right strategy at the right time. But his strategy did create unintended consequences. For instance, to the casual observer the firm was centralized and autocratic. Major as well as minor decisions required George's involvement. To a more seasoned observer, George had instilled an operational culture focused internally on reducing costs and avoiding risk. This culture prevented people from thinking about taking risks that could lead to substantial innovation and revenue growth.

The financial incentives introduced by George reinforced this culture. Workers received incentives tied to overall firm performance and cost reduction targets; there were few incentives that rewarded employees' direct efforts and no incentives were offered to encourage profitable innovation. As a result, revenue growth stalled. Perhaps more pernicious than the slow revenue growth was that most of the employees who were creative and willing to take risks had left the company—and many potential dynamic new hires had chosen not to work for it in the first place. Over time the community came to reflect George's cost-cutting focus and the culture reflected this community. This was the organization Gen had been chosen to lead. Fortunately, Gen had almost a year before taking the reins as CEO to analyze her firm's position and figure out what new direction, if any, WorldCo should take. Even though the company was highly successful, Gen decided that it needed to change; it needed to grow. And not just grow revenue but do so profitably. Three factors drove her to this conclusion.

First, WorldCo wasn't alone in the marketplace. Gen faced a multitude of competitors, several of whom had been reenergized over the preceding couple of years. These competitors were expanding, growing revenues, and encroaching on WorldCo's unique market position. They also had begun to consolidate the industry. It was clear in Gen's mind that focusing on cost reductions alone was not going to keep her company ahead of the competition. In fact, in some ways competitors already threatened WorldCo, once known as the high-tech leader in its field; customers now turned to Gen's competitors for new products, services, and solutions. Customers now perceived WorldCo as one of many distributors in the industry instead of the "hot" industry leader. Gen could see that a day of reckoning—stiff competition and low profit margins—was speeding toward her and she needed to take action.

Second, even though additional cost savings from advancing operational excellence could still be found, it was clear that there was a limit to how much more cost savings she could squeeze out of the operation. The flow of cost reductions clearly was not going to match the torrent of years past. Increasing profit through cost reductions simply could not continue at the same rate indefinitely.

Third, Gen was not alone in her conclusions about the company's future. Wall Street could see the competitive landscape and had come to the same conclusions as Gen. Without new sources of profitable revenue growth, the firm's stock price was going to stagnate.

In many ways the decision to change was not difficult. Given Gen's view of the competitive landscape and the firm's future, the time had come to change strategy, to focus on profitable growth. Gen's new objective was straightforward: change the organization's focus from internal cost reduction to external customer orientation, to seeking out new opportunities to serve customers better, faster, and in new ways that created value for the customer. Doing so would allow WorldCo not only to increase revenue but also to capture some of this newly created value and transform it into profits.

Gen solved one problem—figuring out what strategy to adopt—but in so doing created a much more challenging one: How was she going to implement the strategy to change her company's culture and do so on a global level? She would have to decentralize, delegate, and change reporting relationships, tasks, and financial incentives. Most important, she needed to change the employees' and managers' mind-set—the way they thought, and their expectations about how to succeed at her company. She had to accomplish this difficult task all the while continuing to meet Wall Street's expectations—no easy feat.

She knew that some managers and workers could adapt quickly, but most probably would not adapt easily. And why should they? Things seemed to be going well. The ship was sailing ahead full speed on what appeared to many to be calm seas. Why take the risk of changing course when the current strategy was working well? Even if some in the community could rationally understand her motivation, there were many people whose behavior could not easily be changed. Many people chose to work at Gen's firm because there wasn't much uncertainty in their jobs. How would they respond during the uncertainty of change? How was Gen going to lead them to change their behavior? How was she going to pass her fundamental test of leadership?

William's Challenge

William now found himself the CEO of MandACo, a Fortune 500 manufacturing company that—with more than thirty thousand employees and two-hundred-plus plants—was known for its history of high growth and entrepreneurial spirit.

MandACo holds a unique position in its industry. It is the largest firm in assets and revenues. Throughout the 1970s, 1980s, and 1990s, William's company got its reputation for rapid growth because it actively merged with and acquired many other companies. In fact, MandACo became the largest player on the block by consolidating a large swath of the industry. Back then, customers and competitors were largely geographically local. By consolidating and growing, the firm achieved economies of scale, especially in purchasing, that simply weren't available to smaller local competitors. During this period of consolidation managers had strong financial incentives to be entrepreneurial—and they were.

Each division and each plant was a profit center, meaning that local managers had to manage revenue and costs. These local managers made local decisions to maximize local profitability. Managers who hit their plant's profit target would receive a bonus equal to 40 percent or more of their base salary. These substantial earnings attracted managers who were willing to take risks in order to reap the rewards of success.

This local, decentralized structure meant that each acquisition was easily folded into the firm because it was operated largely independently except for the purchasing of raw materials. Independence meant that managers and employees from the acquired and merged firms often retained their original affiliation and continued to identify with their old company. This process created and reinforced fiefdoms. When promoted, managers did not move around MandACo but stayed in their own "silo." Rarely did a manager from an acquired firm get transferred or promoted to another region or division. A good number of managers had spent their entire career with the same plant, in the same location; the change of ownership was scarcely noticed. Many managers and employees simply continued to identify themselves with their old company that had been acquired. Such promotion and behavior patterns further reinforced the fiefdoms within the company. Initially, such fiefdoms were a good thing: Each boat had floated on its own bottom and the plants and divisions didn't need to coordinate or work together for the umbrella company to be profitable. But now they were holding the company back from evolving in positive ways.

MandACo's consolidation strategy, which was largely funded by taking on more debt, worked extremely well in the growing economy of the 1980s and 1990s, when the mergers and acquisitions led to increased revenues and profits. Although the debt load was great, profits were greater. Economies of scale in purchasing raw materials gave plants advantages over local competitors serving local customers. Capacity utilization was high and investors and managers were rewarded handsomely. Critically, debt repayments were handled easily.

By the turn of the twenty-first century, two new trends fundamentally changed the business environment in which MandACo was competing. First, toward the end of the 1990s and increasingly in the new millennium, customers moved production operations to China, where William's company did not have substantial operations. This shifting of business across the Pacific decreased demand in North America. MandACo, the largest manufacturer in its industry, was the most heavily impacted by this shift in demand. Second, the North American customer profile also changed. Whereas there had been many customers of all sizes, over time two predominant groupings emerged. Customers either remained small "local" customers or became large "national" accounts. These national accounts created commodity markets by auctioning off their demand and playing suppliers off against each other. Only the largest firms could compete for national accounts, and the only way to be profitable with these customers was to be the lowest-cost producer. Some of MandACo's competitors had chosen long ago to focus on catering to these national accounts, making investments and tailoring operations to achieve a low-cost position. Smaller local competitors focused on local customers and differentiated themselves with speed, flexibility, and customization.

MandACo served both national and local customers—often one plant supplied both types of customer. Historically the strategy had been successful, but now the changing environment was leading to critical unanticipated consequences. MandACo's plants were not optimally configured to serve either the large national customers or the local customers. Many of its plants were high-cost producers for national customers and couldn't provide the speed, flexibility, and customization that local competitors could provide. The shift in demand had left MandACo stuck in the middle—a classic business predicament described by Michael Porter in *Competitive Strategy: Techniques for Analyzing Industries and Competitors*.[4] MandACo wasn't set up to serve either national accounts or local accounts profitably.

Making matters worse, the decrease in demand meant that many plants were sometimes idle, and it wasn't long before plant managers—blessed with entrepreneurial spirit and aware of incentives that rewarded the strongest profit centers—began competing against each other to capture business. It is one thing to drive prices down to compete effectively with other firms; it is quite another to drive down prices because of competition among divisions within your own company. This internal competition amplified problems because plants would take on any business—even business that did not fit the plant's capacity, which further raised production costs.

The changing environment, refocused competitors, and internal competition evaporated profits. By the time William became CEO, MandACo's stock price had dropped by 30 percent. Plants and divisions were refusing to cooperate with one another and were blaming one another for the company's problems. His first response was to reduce capacity by shutting down manufacturing facilities, but demand shrank faster than he could reduce capacity. With little financial room to maneuver and still needing to make debt payments, he was unable to invest in and update production equipment. His plants and equipment had become outdated.

When it became clear that the old cost-cutting paradigm would not work to revive the company, he launched a new strategy: he decided to reorganize the company to match specific plants and divisions to the different types of customers, and to focus on operational efficiency. One portion of the company would serve national accounts and the other would focus on local customers. Achieving a low-cost position to serve national accounts would require massive investments in equipment; to raise the capital needed to make these investments, he would have to sell a very profitable division of the company.

Some aspects of William's strategy were easy to implement. Selling a division, buying equipment, and paying down debt are transactional and easily accomplished. William's real challenge lay in changing his organizational structure and changing the company's culture so that managers would stop making decisions according to what they thought would be good for their individual plants and divisions and start making decisions that benefited the whole company. Making these changes required centralizing the decisionmaking process, reorienting tasks, and changing financial incentives to encourage sacrifice and cooperation instead of entrepreneurial spirit and competition. William had to orchestrate all of these changes while meeting his debt repayment schedule to keep the company out of bankruptcy.

Unlike Gen's situation, where there was a great deal of resistance to change, William knew that even though some managers would be slow to change, many managers were ready to embrace change. These managers knew the company was in trouble and were willing to take risks and change their behavior to save it—once they knew how. Still, William realized that changing the behavior of his community would be like herding cats because of years of hiring people with an entrepreneurial spirit. Changing the organization would not be easy. How was William going to pass his fundamental test of leadership?

3 Why Leading Change Is So Difficult

┌─ KEY POINTS ─────────────────────────────────────┐

➥ The fear of losing something personal (such as position, income, status, ego, friendships, identity, reputation) during organizational change creates resistance to change.

➥ Fear is amplified as uncertainty about the change and what it means to individuals within the community increases.

➥ The leadership challenge of enabling organizational change requires overcoming the community's fear and uncertainty about the change.

└──┘

Academics have studied organizational change for decades and have developed numerous processes for leading change. The vast majority of the processes discussed in the literature refer in some way to the notion of unfreezing expectations, changing them, and refreezing them again.[5] Every business student at both the undergraduate and graduate level takes classes on how to lead organizational change. All management consulting firms—whether focusing on strategy, information technology, supply chains, or operational effectiveness—offer change management and implementation services and collectively earn hundreds of millions of dollars or more each year for their guidance. Scores of popular business books provide the basic steps for leading change. All of us are connected to various types of organizations, have observed others leading change or have led change efforts ourselves, and have learned from the successes—and failures—of these change efforts. Why, then, with all this available knowledge, experience, and learning have studies shown that 70 to 90 percent of change efforts fail?[6]

Barriers to Organizational Change

There are myriad well-known reasons why people fail in their attempts to change organizations: The choice of strategy was poor. People naturally resist change. Some community members found it in their best interest to actively undermine the leader. Incentives and organizational structures were not aligned to support the change effort. The leader was not focused on the change effort or did not get buy-in from the community. Senior management did not commit to the change. Too much change was attempted at once. Community members received insufficient training for their new roles. Although a list of more minor reasons for the failures of organizational change could continue, there is another substantial reason that is seldom acknowledged.

Fear as a Barrier to Organizational Change

Underlying many of the aforementioned reasons for the failure of organizational change initiatives is fear of losing something valuable. Fear is an emotion—perhaps one of the strongest of human emotions that community members experience when they perceive there is a chance of losing something personally valuable. When community members become fearful they act to protect themselves, and this usually is not in the leader's or the organization's best interest. Fear can cause your community to resist and undermine organizational change. The fear of organizational change comes from a combination of two forces: fear of loss, and uncertainty.

Fear of Loss

Community members often perceive that they could lose something personally valuable during organizational change. Position, income, and status as well as ego, friendships, identity, and reputation all are valuable—very valuable—possessions that are intertwined with a person's job and community and identity. These potential losses can be especially great for middle managers, whose position, income, and status might diminish much more than those of others in the firm should they get moved into another role or forced to leave the firm. Fear of loss can readily translate into anger when such losses occur, which can only further undermine organizational change. Middle managers represent great value to an organization because of their specific knowledge of the community and deep relationships within it, and it is exactly this value that may be largely lost if they are moved to another position or are forced to find employment elsewhere. The value of such organization-specific knowledge and community-specific relationships—which is the person's unique human capital—does not transfer easily to another organization and may be lost. Members of the community who possess more specialized knowledge and skills may be better able to switch jobs or firms with a smaller loss of value, as long as their specialization is in demand in the marketplace. Organizational change can instill fear in everyone touched by it, but fear among middle managers may be greater than for others in the community because their perceived potential for personal loss is greater.

Fear of losing all of these possessions is elevated during organizational change, especially when the organization is stagnant or shrinking. In these circumstances, organizational change increases the likelihood that community members will be forced to take a job lower in the hierarchy, with less income and reduced status. Such a change can damage egos, endanger friendships, force new and undesirable identities on the demoted, and dilute reputations. Fear quickly turns to anger when any of these losses are realized.

Leading change is far easier when your organization is doing well and growing than when it is stagnant and shrinking. Rapidly growing organizations face fewer impediments to change because growth, if managed wisely, can bring new opportunities that offer an increase in these kinds of human capital and in their value, instead of diminishing them, so that change promises positive outcomes for the existing community—especially its middle managers. Ultimately, the perception that personally valuable possessions can be lost is what induces strong emotional reactions of fear and resistance to change.

Uncertainty as an Amplifier of Fear

Uncertainty, especially when it increases the perceived likelihood that personally valuable possessions might be lost, amplifies the fear generated by the potential for loss. Every organizational change initially brings with it much uncertainty. Community members often are uncertain about many things: What will the organization look like in the future? What opportunities will be available? What will individuals need to do to avoid failing, let alone to succeed? What personal capital and valuable possessions might be lost as a result of the change? What additional tasks might be forced on individuals?

Academics often use what is called an "economic" definition of uncertainty: the idea that a range of outcomes is known but the probability of outcomes cannot be estimated. The definition of uncertainty adopted in *Leading Change* differs from academic definitions because not only the probabilities but also the alternatives are unknown. Uncertainty in this book refers to not knowing what the future will bring.[7]

Uncertainty by itself does not instill fear. Every day our lives are filled with uncertainty. Will traffic get snarled today and delay our arrival at work? Will our flight leave on time or be delayed by bad weather? Will we win the lottery? These uncertainties do not create fear or cause us to resist change. In fact, the opposite is true—we live comfortably with uncertainty and even sometimes seek it out. We leave for work, arrive at the airport, and occasionally purchase lottery tickets and quickly adjust should we experience an unfortunate (or unlucky) outcome. We don't resist or avoid these activities simply because of uncertainty. Yet when specific kinds of uncertainties, especially those involving changes at work, are not quickly eliminated they can spark behaviors that rapidly intensify to create tremendous resistance to change.

The uncertainty surrounding organizational change is an especially perilous concern because people do not treat the likelihood of various outcomes the same. It is well known in academic circles that people tend to fear the loss of something valuable more than they care about receiving future rewards.[8] People concentrate on avoiding the possibility of losing something they already possess. For instance, it is common for people to hold on to stock or real estate when the market price falls below the original purchase price, even when they expect the price to drop further. The financial crisis of 2008–09 provides many illustrations of this behavior. Such investors fear locking in the financial loss from selling and consequently hold on to such investments even though they expect that prices will decline further.

People fear the loss of something valuable they already possess and often will go to extreme lengths to protect it.[9] Because of this mind-set, few want to be the first to adopt the organizational change for fear that if they do the wrong thing, adopt the wrong behavior, or are not successful, they will be punished in some way or fired. Few want to give feedback to senior managers or ask questions of them for fear that they will be branded troublemakers or will stick out like nails only to be hammered down. Few want to adopt the new behavior for fear that doing so increases pressures on friends to change, which in turn creates social pressure for them to conform to the group. Even in organizations that encourage dissent and that have individuals with strong personalities, fear of personal loss can motivate managers and workers alike to keep their heads down, follow the herd, and resist change.

Worse still, uncertainty coupled with a perception of potential personal loss invites rumors. Community members seek out information that will help them reduce uncertainty around the bad outcomes and to manage their risk of personal loss. Experiencing uncertainty about the potential for personal loss, community members naturally start, pass along, and pay attention to rumors because it gives them at least some type of information, albeit often erroneous information, to help them manage their anxiety. Unfortunately, people are attracted especially to rumors that stoke their worst fears and pay less attention to positive rumors—rumors that suggest change will make them better off—because it is the negative rumors that spotlight outcomes they want to avoid. These negative rumors act to increase the perceived likelihood of personal loss, thereby amplifying fear, which further erodes and ultimately paralyzes change efforts.

How Fear Arrests Change

To illustrate the role of fear in arresting organizational change, consider Gen's organization immediately after her initial announcement about changing the organization. Gen announced to her community that after many years of a successful internally focused cost-reduction strategy, she planned to decentralize the firm. Decentralization would shift the focus to creating new services, new products, and new value for customers, which she hoped would trigger profitable organic growth. She launched her change effort by making presentations and giving speeches to the firm's senior leadership, telling her board of directors and the investor community about her strategy, and announcing the change in her company's newsletter.

Most of Gen's managers did not understand just what the message meant for the organization or for them personally. Few of the senior managers had been on the team that determined the new strategy, so most of the community did not understand the reason for the change. WorldCo had done very well over the preceding five years and most thought it could continue its success by sticking to the same path. Many in Gen's community wondered why the company couldn't continue to do well with the old strategy. Even if they did understand the need for change, community members were uncertain about what specific behaviors Gen wanted them to adopt, which created fear. What would decentralization or creating value for the customer mean in terms of changes in day-to-day activities for community members? Few understood how their roles would change. What would they have to do to keep from failing in the new organization? Many, especially those who had focused on driving down costs, were afraid that their skills and knowledge would be devalued under this new strategy or, worse, that they would be no longer needed. Uncertainty coupled with a perception of personal loss generated a fear that worked against Gen's announced changes.

William faced similar reactions from his community after his initial announcement about changing MandACo's strategy. William announced that he was going to break down fiefdoms by centralizing around several functional activities: manufacturing, marketing, and sales. Doing so would allow him to coordinate the company's investments in new plants: some would specialize in high-volume production to serve large national accounts and others would be reconfigured to flexibly serve local accounts. Centralization of marketing and selling would eliminate competition among plants and allow for the efficient matching of customer needs with plant capabilities. William also announced that funding these changes might require selling some plants, closing others, and reconfiguring practically the entire firm.

William communicated his new plan to analysts, sent memos and announcements to his workforce, and met with key managers to present his changes. Yet few within his community understood how the measures outlined would change the organization. Middle managers in particular did not understand what this new strategy meant for them and their day-to-day activities. Would their operation be sold? If not, how would they fit into the new organization? Would they lose their jobs? If they kept their jobs did they have enough of the right kind of knowledge and skills to succeed in the new organization? These middle managers were uncertain about what the future might bring and perceived that there was a good chance that they would incur personal losses.

Most organizational change efforts face similar responses from those whose behaviors and decisions the leader wants to change. Communities are fearful of change especially when they perceive much uncertainty about the likelihood of personal loss. The leadership challenge for Gen and William could be redefined: How can they overcome fear and uncertainty in their communities to enable organizational change?

Summary

The organizational inertia that resists and slows down organizational change efforts comes from many sources. Chapter 3 identified what may be a common underpinning for inertia, the fear of losing something of personal value, such as one's position, income, status, ego, friendships, identity, and reputation. Fear causes people to act in their self-interest, which commonly involves resistance to organizational change. Fear is perhaps most prevalent among midlevel managers whose unique professional assets may not be as redeployable to other positions and other firms as those of other types of employees. Uncertainty about organizational change and what it means to individuals within the community amplifies fear. Fear is converted to anger, should loss occur, and this in turn further increases resistance to change. Your leadership challenge for enabling change is to overcome your community's fear and uncertainty about what the change will mean to its members.

4

Enabling Organizational Change

―――― **KEY POINTS** ――――

⇥ Leading change requires overcoming your community's uncertainty and fear.

⇥ Organizational change can be led using either a compliance or a commitment approach.

⇥ The compliance approach relies on measurement and reward or punishment to achieve desired behavior.

⇥ The commitment approach relies on developing trust and understanding to achieve the desired behavior.

⇥ Trustworthiness is based on your community's view of your character, goodwill, and ability.

⇥ Building trust and creating understanding call upon a leader to follow three principles:

　—Converse with the members of your community all at once.

　—Protect your community's conversation.

　—Demonstrate that you listen to your community.

I f fear and uncertainty are forces that poison and arrest organizational change, what forces breathe life into it? If you are like most leaders, you probably have been educated to manage organizational change and may be aware of many principles and approaches for doing so. Get the incentives right, build a guiding coalition, unfreeze expectations, adjust many organizational elements at once, get some early and easy wins, punish those who do not comply—these are just some of the prescriptions offered by the scores of managing-change frameworks available. Yet isn't it precisely the application of these management frameworks that generates the 70 to 90 percent failure rates of change initiatives? These frameworks are necessary and useful, but something still seems to be missing. What principles can you add to your repertoire to increase the likelihood of successfully enabling change in your community?

Two Ways to Enable Change

Most approaches to leading organizational change fall into one of two broad categories, which I call "compliance" and "commitment."[10] Both approaches can enable change successfully in the appropriate circumstances, but they use drastically different means for doing so. Compliance relies on coercion or inducement to get your community to comply with your change initiative. The commitment approach entails seeking to gain the commitment of your community to enact your change initiative but also requires particular commitments from you. Although this chapter describes elements of the compliance approach, the chapter's main focus is on the commitment approach because it is this latter perspective on which *ChangeCasting* is built.

The Compliance Approach

In the compliance approach to enabling organizational change, management gives directives, specific actions by subordinates are required, and managers then measure or evaluate those efforts or outcomes to ensure that workers comply with the directives. Compliance most often is achieved by means of two types of managerial levers that influence worker behavior by stimulating emotions that overcome the fear and uncertainty that cause resistance to change. The first managerial lever is called "measure and punish," and also "command and control." The exercise of authority and the fear of punishment are used to overcome resistance to change. The Prussian emperor Frederick the Great said, "Soldiers should fear their officers more than all the dangers to which they are exposed. ... Goodwill can never induce the common soldier to stand up to such dangers; he will only do so through fear."[11] Frederick the Great's approach to overcoming resistance is the archetypal use of power to command and control and force compliance. With such a power approach, the fear and uncertainty of organizational change is overcome by a greater fear of being detected as having not followed orders and the expectation of a severe punishment. In essence, one fear is used to trump and overcome another.

The second managerial lever used to compel compliance is measure-and-reward. Instead of using fear to control others, leaders use positive inducements to generate different emotions, the most common being greed, to overcome fear and uncertainty. Measure-and-reward managers shape behavior by offering strong financial inducements that are received contingent on following commands or achieving measurable goals and objectives. Financial inducements have become standard levers in the modern management toolbox. Today, practically every for-profit organization offers some type of contingent pay, stock options, bonuses, or other financial inducements tied to achieving specific goals and objectives to stimulate effort and lead changes that can drive increased revenue and profits. The importance of this category to compliance is confirmed by the existence of an entire subfield of economics that is dedicated to studying the structure and magnitude of financial inducements for both managers and workers, so that supposedly optimal inducements can be offered.[12]

It is not unusual for managers to use both compliance approaches—the carrot and the stick—to enable organizational change. The Ford Motor Company in the early twentieth century provides an excellent example of the use of compliance to manage organizational change.[13] Henry Ford changed the industrial landscape of America through his approach to change management. He introduced a stunning organizational change in the automotive industry by moving from craft production to vertically integrated mass production on an assembly line. His was one of the most vertically integrated mass-production enterprises the world has ever known; small wonder that *Fortune* magazine in 1999 named him the Businessman of the Century. His new management system even acquired a special name, Fordism.

Workers, however, did not enjoy Ford's mass-production system, for it was deskilling and dehumanizing. Massive employee turnover (380 percent in 1913, during the heart of his change initiative), high absenteeism (upwards of 10 percent per day), and minimal effort characterized worker responses to the introduction of Fordism, responses that greatly disrupted Ford's manufacturing productivity. To eliminate these problems, in 1914 Henry Ford launched a powerful compliance initiative by more than doubling workers' pay compared to what other firms paid. Four months after opening his Highland Park, Illinois, plant he offered a daily wage of five dollars as an inducement to increase workers' willingness to withstand the drudgery and burden of keeping his assembly line moving. But there was a hitch: workers received half of this pay with each paycheck and the other half as a large balloon payment after the passage of one year if, and only if, he or she was still employed with Ford at that time. A worker who did not comply with all of Ford's directives was quickly fired and did not receive the other half of the pay being held back by Ford.

In addition to the directives imposed on them at the factory, workers were admonished to change their behavior even at home by adhering to certain "character" requirements. These character requirements were monitored by Ford's sociology department, which would periodically visit worker homes and neighbors to monitor adherence. Failure to comply with character requirements, just like failure to comply with work requirements, led to termination.[14]

The inducement of receiving such a relatively large wage and the fear of losing half of one's accumulated pay with no similar financial opportunity with other employers coerced many to comply with Ford's directives. Indeed, Ford's five-dollar-a-day wage offered practically one of the only opportunities for workers to rise above the squalid conditions endured by laborers in that era. This financial incentive, which involved elements of both measure-and-punish and measure-and-reward, not only gave Ford substantial control over his workers that forced compliance but also was an extremely profitable policy for his company.

Using control to enable organizational change is as functional in the early twenty-first century as it was in the early twentieth century. Today, most change efforts involve some elements of these compliance approaches. For instance, early on in their change efforts, both Gen and William introduced new compensation and bonus structures to reflect their new priorities. These measure-and-reward structures not only signaled new, desired outcomes but also measured individual and organizational performance against stated goals and, in some instances, actions. By aligning financial rewards with the new organizational objectives, Gen and William expected to induce managers and the rest of their communities to embrace the change initiatives. Both organizations also used the measure-and-punish approach. Some managers ultimately were let go because they refused to enable some changes. Others were reassigned to new tasks with the explicit expectation that future career progression depended on achieving the stated goals. In both companies, several managers eventually were punished with termination for not achieving their assigned goals.

Using the levers of compliance can be effective, and conventional wisdom suggests their use is necessary, at least to some degree. In some instances, the emotions that compliance engenders are sufficiently strong to overcome the fear and uncertainty that typically cause resistance to organizational change. In other words, the compliance approach uses one fear (or desire) to overcome another. Generating such strong emotions offers many benefits for leading change. Commands can be implemented quickly. The community's attention is focused on accomplishing specific directives and measured outcomes. And following commands can make an organization highly responsive, especially when the leader provides clarity and communicates specific actions the community should take to enact the change. Military organizations must rely on this type of chain of command, especially in combat.

As you might imagine, a heavy reliance on the compliance approach for leading organizational change also entails some challenges and costs. Perhaps the most obvious one is that the breadth and depth of typical local labor markets at least in developed countries make it easier today than ever before for workers to escape coercive forces like the ones used by Ford a century ago and simply look for other employment. Similarly, modern labor laws, at least those in the world's advanced economies, and the increasing use of social approbation against global corporations perceived as using physical coercion against their workers make it difficult for many companies to use coercion. Of course, this is not true in all situations, especially in less-developed economies.

Another problem with the measure-and-punish approach is the type of behavior it induces. Workers focus their effort and behavior on the bottom line of what is required and measured, doing just enough to avoid punishment. Certainly the potential loss of earned income successfully coerced many of Ford's workers to comply with his dictates, but only obeyed with their hands, not their minds and hearts. They followed commands to the letter, but were only compliant—not inspired.

If you do not know in advance precisely what workers need to be doing to effect your change effort or if you find it difficult to measure a few desired outcomes and the effort they require, your organizational change initiative likely will not proceed well if you rely strictly on a compliance approach. Individuals will adhere to the directives whose results are measured and will wait for further commands rather than taking the initiative to determine for themselves what needs to be done to help the organization succeed. Instead of creating value for the organization, they may focus on political maneuvering to avoid punishment and on getting others punished so that they can move ahead—the so-called blame game. Replacing one source of fear with another still results in people being driven by fear, which may narrow the range of behaviors workers willingly engage in to benefit your organization or company.

The measure-and-reward approach comes with a different cost. Successful organizational change may require many different tasks to be performed, some of them easy for management to measure and others that are difficult to measure or can't be measured. For instance, Ford could readily measure that workers were present at their work stations and engaged in their tasks because a multitude of supervisors scoured the assembly line watching for slackers. Yet, the quality of their performance could not be measured easily, which resulted in quality problems for Model T's coming off the assembly line. Bolts were not always tightened and screws not fully threaded. It was not uncommon for a part to fall off a car after it left the factory, arrived at a dealership, or was en route to the new owner's home.[15] A focus on measurement may encourage workers to shirk on those outcomes desired but not measured.

Most workers are assigned multiple tasks, which may lead to workers' focusing their energies on those tasks that are measured to the detriment of tasks that are not measured, or at least not measured well.[16] A related and pressing issue today is the desire to achieve objectives such as profitable growth but without compromising ethics. Profits are easily measured, but ethical behavior is not, especially in real time. High-powered financial inducements cause workers to focus on what is being measured and often to siphon effort away from behaviors and outcomes not easily measured to those that are. To a large extent the roots of the 2008 financial crisis can be found in the easy measurement of revenues and profits and the difficulty of measuring ethical behavior.

It is precisely those instances in which managers offer high-powered incentives to increase profitability, difficult-to-measure ethical behavior may be sacrificed. In this case, the old saw that "you get what you pay for" should include the additional caveat that "you don't get what you don't measure." Ultimately, with a compliance approach, you only get what you measure and pay for, and over the long run you shouldn't expect anything else.

The Commitment Approach

The commitment approach to enabling organizational change represents an alternative set of managerial levers designed to overcome the fear and uncertainty that cause resistance to change. By commitment we mean stimulating your community to make a commitment to enacting organizational change. In so doing you capture minds and hearts as well as hands. How can you stimulate such a commitment?

If fear and uncertainty poison and arrest organizational change, then building trust and creating understanding provide antidotes to enable it. If your community trusts you, it believes in your character, goodwill, and ability to guide it through change. This trust, if strong enough, can diminish fears of personal loss for many in your community because they believe you will do what you say, you are concerned about the welfare of your community, and you possess the ability to succeed. If your community understands why change is needed and what they must specifically do to change their behavior to make their organization successful, then the uncertainty that amplifies fear can greatly diminish. Thus, trust and understanding can reduce the perceived likelihood of personal loss and uncertainty that stalls change efforts.

Building trust and creating understanding may sound intuitively desirable, but how do you do it? Most research on building trust assumes that you either have it or you don't, instead of describing how to build or repair it.[17] Most research on understanding largely assumes that you create understanding by the way you shape your message and what you communicate. But the reception of the message is affected by different ways of listening. Listeners often are viewed as passive: you, not they, create understanding, even though, as you know, creating understanding must involve the active participation of the whole community.[18]

Three principles can help you generate commitment that overcomes resistance:

1. Converse with all the members of your community all at once.
2. Protect your community's conversation.
3. Demonstrate that you listen to all members of your community.

These three principles form the basis of how you and the other community members can make reciprocal commitments that build trust and create understanding to enable change.

Principle 1: Converse with the Whole Community All at Once

Aristotle contended that people trust a leader whom they perceive to have integrity, beneficence, and capability. In *Rhetoric*, he stated:

> There are three things which inspire confidence in the orator's own character—the three, namely, that induce us to believe a thing apart from any proof of it: good sense, good moral character, and goodwill. False statements and bad advice are due to one or more of the following three causes. Men either form a false opinion through want of good sense; or they form a true opinion, but because of their moral badness do not say what they really think; or finally, they are both sensible and upright, but not well disposed to their hearers, and may fail in consequence to recommend what they know to be the best course. These are the only possible cases. It follows that any one who is thought to have all three of these good qualities will inspire trust in his audience. The way to make ourselves thought to be sensible and morally good must be gathered from the analysis of goodness already given: the way to establish your own goodness is the same as the way to establish that of others.[19]

Aristotle's triple for creating a perception of trustworthiness—good sense, good moral character, and goodwill—is commonly described as integrity, beneficence, and capability. Modern research supports this view. Scholars have found that individuals who are perceived to possess character, goodwill, and ability are viewed as trustworthy.[20] If these perceptions form the bases of trust, what can you do to generate such perceptions?

A body of communications research shows that the most important factor influencing what a community member thinks of you is what other community members think of you.[21] In other words, the extent to which one person thinks you are trustworthy depends on the extent to which his or her friends and associates believe that you are trustworthy. Understanding this relationship among community members is vital for thinking about how you as a leader can build trust. It implies that if you communicate your message to one person or one group at a time, these people will soon speak with friends and colleagues to make sense of your communication. If these peers and friends either have not heard your message, which is likely if you are speaking to people sequentially, or don't believe in it (or you), then they can and probably will weaken your message by providing incorrect interpretations to others, or undermine it with false rumors. At a minimum, hearing different messages from their friends and colleagues as compared to your messages creates confusion for them about what to believe. The confusion reduces understanding and increases uncertainty instead of attenuating it.

Speaking with your community one member or one small group at a time does little to enable change because doing so creates space and time for the forces of fear and uncertainty to work against you via rumors and secondhand reports. Unless the entire community already views you as trustworthy, you will be sailing in light winds against a strong current of confusion, uncertainty, and fear.

One way to slow and reverse this current is to speak with your entire community at the same time instead of sequentially. Then everyone gets the same message at the same time. Such parallel communication can diminish rumors, although not eliminate them. It can increase the likelihood that everyone can perceive your character, goodwill, and ability, which will encourage them to discuss your message with each other and reinforce your initiative instead of undermining it.

Conversing with your whole community simultaneously is a necessary first step to establishing trust, to be followed up with two others.

Principle 2: Protect Your Community's Conversation

In teaching students of all ages, especially executives, I have learned that nothing shuts down a conversation and limits learning more than the professor's failure to protect the conversation. For instance, letting a few individuals consume a lot of air time and dominate the conversation; allowing put-downs, wisecracks, and jibes; allowing some to make jokes at the expense of others—any communications that tend to stifle the conversation or discourage some from participating assault the conversation.

Protecting the conversation is the gateway to engaging in an open, candid, and meaningful dialogue. It is vital because good conversations can lead to change in people's thinking and behavior. If conversation is not protected, conflict and negative emotions block learning and arrest change. Another consequence of failing to protect the conversation is that few are willing to participate, let alone try to understand one another, because they fear they will be attacked, hung out to dry, or have other unpleasant experiences. In an educational setting, students will not engage with the material and learn if the conversation is not protected.

The same is true in organizations—with one twist. In addition to avoiding dominance, putdowns and jokes at others' expense, leaders also must restrain themselves from blaming or punishing those who speak up. Sophocles, and later Shakespeare, noted the importance of protecting conversations by not "shooting the messenger"—blaming a person for saying something you do not want to hear. Conversations elicit insightful information as to what your community fears and what uncertainties they perceive. Without finding a way to credibly commit to protecting your conversation, and the messenger, you will never find out what is getting in the way of your change efforts.

Protecting the conversations you have with your community means that it is up to you not only to keep from shooting the messenger but also to protect the messenger from being shot by others. Punishing, reacting negatively, and using your position to publicly point a finger at a messenger guarantees that this person will not trust you, will not invest in developing understanding, and likely will never again confide in you. Neither will anyone else. Your community will instinctively learn to keep its collective head down and remain silent.

In contrast to this, by developing a reputation for unequivocally protecting the conversation you will access much of the information you need to enable organizational change. Without developing such a reputation, rumors, misinformation, false information, and misunderstandings based on fear and uncertainty will not get disclosed to you and will undermine your change effort.

Protecting the conversation will allow you to learn about your community's fears and uncertainties, so that you can thoughtfully react to reduce uncertainty and fear. Protecting the conversation also means being open to others' communications and messages, even when they are critical. Listen to the message and evaluate it on its merits. Even if it is critical, express appreciation for the contribution. Knowing what rumors, misinformation, and misunderstandings are circulating gives you a chance to set the record straight and find new messages and new ways of communicating with your community to build trust and create understanding.

Principle 3: Demonstrate That You Are Listening

"He just doesn't listen!" How many times have you described your boss this way? Could people say this about you? This one phrase can open the floodgates to great emotional weight that can swamp your change initiative. If your community perceives that you are not listening, not considering the insights they provide and the information they offer, you are demonstrating that you don't respect them.

Demonstrating that you listen to your community is a positive step in building trust and creating understanding and showing that you respect your community. Respect is an important element of trustworthiness. Not listening calls into question your character, goodwill, and ability, the three elements of trustworthiness. If you don't demonstrate your respect for your community, how can your community respect and trust you? If you don't want to listen to and understand your community why should they listen to and try to understand you? Instead, they will listen to peers and friends who will likely have very different messages that will unleash negative emotions. They will compound your community's fear and uncertainty and slow your change effort.

Listening to your community, though, may be more difficult than you imagine. Recent research indicates that as a leader you may hold biases that make it difficult for you to listen to your community. Studies have indicated that holding a powerful position leads managers to consider only their own vantage point and experience and to adjust insufficiently to their community's perspectives.[22] Making matters worse is that if your community perceives that you do not listen, they may conclude that you are likely to use compliance instead of commitment to advance organizational change. Such a perception not only undermines your trustworthiness but also may cause your community to resist change even more, until you feel compelled to use a compliance approach, which becomes a self-fulfilling prophecy for the community. Either way, your community's uncertainty and fear will increase if it perceives that you do not listen to them.

There are many ways to demonstrate that you listen. Learn some active listening skills. Look the speaker in the eye, paraphrase what the speaker has said, validate the speaker's position verbally and with gestures, listen for the expression of feelings and describe them. These are skills you may have heard about, learned, and practiced. All of these techniques demonstrate that you are listening to someone speaking, understanding them, and respecting them.

Using listening skills is not the only way to demonstrate that you listened. An even more important indicator that you are listening is recounting in public forums what others have said; this clearly signals to the broader community that you have been listening. Stating not only the questions raised but explaining your responses to them further advances the perception that you listen to and understand your community. Showing specific outcomes, invest-ments in the organization, organizational changes that illustrate that you listened all can provide powerful signals to your commu-nity that you are a good listener. Of course, it may be difficult to have these signals at the ready while conversing face to face with someone. Nonetheless, practicing good listening skills and giving these other signals demonstrate the respect for your community that enables you to build trust and create understanding.

Enabling Change

Conversing with your whole community simultaneously, protect-ing your community's conversation, and demonstrating that you listen to your community are the practices that create the precon-ditions for using a commitment approach to leading change. These process steps seek commitments from your community but also require commitments from you. In short, if you commit to and execute these three practices you will begin replacing fear with trust and replacing uncertainty with understanding, which will enable organizational change.

The early-twentieth-century Ford Motor Company offers a well-known example of use of a compliance approach for managing organizational change; the late-twentieth-century Ford provides a favorite example for using a commitment approach to managing change.[23] Ford's financial difficulties began in the late 1970s. By the end of the double-dip recession in President Ronald Reagan's first term (1980–82), Ford had lost more money than any other company in the history of the modern industrial world—some $2 billion. Nine production facilities had been closed, sealing the fate of company towns in which they were located. The situation was dire.

Ford's new management team of Philip Caldwell, Donald Petersen, and Harold Poling launched a change effort to dig the company out of its financial crisis. The change initiative came to be known as "Quality Is Job One" (you may remember some of the TV commercials that aired in the 1980s). And the job was a big one. These executives discovered, from surging Japanese competitors, a new way to manufacture cars. They learned about the ideas of Edwards Deming, quality circles, and *kaizen*, which is the constant improvement of quality.[24] From Japanese manufacturers Ford executives learned that "not machines [but] the people . . . made the difference in quality and productivity."[25] Adopting this new mind-set, where workers made suggestions on new car designs and continually improved the production of existing designs without dictates from management, was a monumental departure from Ford's historical modus operandi.

For nearly seventy years Fordism had reigned, creating deep divisions between unions and management. Steve Yokich, the vice president of the United Auto Workers (UAW), asserted that management believed that "workers have no brains . . . just legs, arms, and hands."[26] Workers for their part considered managers uncaring and callous. Empowering the workforce and depowering supervisors and management was such a radical and unfamiliar concept that initially neither workers nor managers understood it.[27]

Caldwell, Petersen, and Poling nonetheless launched a change initiative to reorient the entire organization. They chose a commitment approach instead of a compliance one. They began by launching a conversation with UAW leaders. During the conversation both sides chose to listen, chose to protect the conversation, and in the end they came up with an initial groundbreaking agreement in 1979. The Ford and union leaders worked together to implement and support a new process of involving employees. After signing the agreement, management and UAW leaders together visited every Ford facility as quickly as possible and jointly spoke to workers. They launched conversations with workers and managers about the goals of the organization and about the change initiative.

As might be expected, workers and managers were suspicious because of almost seventy years of distrust. Yet the leaders listened to their community. They made a number of adjustments to their change initiative in response to what they heard. For instance, the employee involvement process was kept distinct from the suggestion program because workers viewed the latter as a management platform, not their own program.

Leadership learned that a parallel program to help managers learn how to be coaches instead of authorities was needed. At no time was anyone punished for providing feedback. Rather, the leaders made commitments to workers such as asking them voluntarily to join quality circles, which involved working on solving problems of their own choosing without needing approval from management to do so. Management did not measure outcomes of these projects and therefore did not focus on compliance. Many adjustments were made in response to conversations indicating not only that management listened to its community but also that it protected the conversation.

Slowly, the change initiative began to take hold. Initially, only four out of sixty plant managers agreed to voluntarily adopt the initiative, but this was enough for small successes to begin to appear. Workers began to find and solve problems on their own that had a material impact on the firm. The leaders added these examples to the conversation, thereby demonstrating the successes that were possible by adoption of the change initiative. As the conversation continued and additional successes emerged, the new approach to operations began to take root more broadly. Although there was no official end to the initiative, it was clear that it took perhaps five years from the initiative's launch for the change to reach most parts of the company.

As you might imagine, the leadership challenge was immense, and involved many small steps along the way.[28] From my perspective, none of these steps would have occurred if the leadership team had not launched a commitment process to build trust and create understanding. Conversing with their community, protecting the conversation, demonstrating that they listened—all clearly were present in Caldwell, Petersen, and Poling's approach to leading change.

Most commentators conclude that Ford's change process was successful. Indeed, the success of the 1980s change process ultimately may have contributed to Ford's ability to weather the 2008 financial crisis. In contrast, General Motors typically used a compliance approach for leading change, which may not have served it well during the recent economic crisis. Ford's leaders did enable change. Yet, the change process took many years, was very costly, and the company was in deep crisis before the change efforts took hold.

This chapter has presented three principles that enable change. Yet the question for twenty-first-century leadership is not just how to enable change, but how to accelerate it. The competitive global environment in the twenty-first century means that organizations in crisis today likely wouldn't survive the five years it took Ford to pull itself out of the hole. And not only organizations in crisis need to change. Those that are out in front of competitors and want to maintain their lead must also be able to adapt faster than their competition. Thus, all companies require a method to accelerate change.

How can you accelerate change in your community? The next chapter introduces a new methodology for accelerating change by using Web 2.0 technology. I refer to the combination of this technology with the principles discussed in this chapter as Web 2.1.

Summary

Research has shown that 70 to 90 percent of organizational change efforts fail to meet expectations. Approaches to organizational change can be classified as a compliance approach or a commitment approach. Although each approach can enable change, Change-Casting relies on the commitment approach. Enabling change through the commitment approach requires application of three practices: communicating to your entire community simultaneously, protecting your community's conversation, and demonstrating that you are listening to your community. These can enable you to capture the hearts and minds, as well as the hands, of your community in support of your change effort.

5 Accelerating Change in a Web 2.1 World

━━━━━ KEY POINTS ━━━━━

➥ Web 2.0 technology offers new ways to converse with your entire community all at once, protect the conversation, and demonstrate that you have listened.

➥ ChangeCasting is a Web 2.1 three-step process:

1. Sending digital videos of your messages to your community.

2. Enabling anonymous web-enabled feedback to protect the conversation.

3. Discovering common concerns and visually demonstrating that you have listened to your community.

The term "Web 2.0 technology" refers to new web-based technologies that facilitate communication and secure information sharing, interoperability, and collaboration on the World Wide Web. This somewhat amorphous term comprises, among other things, web-based communities, hosted services, and applications such as social networking sites, video sharing sites, wikis, blogs, vlogs, and so forth.

The essential feature of Web 2.0 technologies is that they open up new means by which people can communicate with each other.

Advances in Web 2.0 technologies, computers, and information technology in combination potentially provide new solutions for conversing with your community to enable and also accelerate change.

These technologies can be used in many ways and for many purposes. They do not necessarily imply a process or a set of guidelines for generating the kind of conversation you need to be successful at leading and accelerating organizational change. But their availability is a precondition for the conversations you want to generate. So you need to know what they are and how best to use them to lead change.

In this chapter I describe some of the more important Web 2.0 technologies and introduce a process whereby you can use these technologies to lead and accelerate change. These technologies, if applied with an appropriately designed process, can help you hold a useful and valuable conversation with your community and accelerate organizational change. The process introduced here, what I call ChangeCasting, has three structured process steps, which together constitute a new way to implement the three principles for leading change by generating commitment. (Additional guidelines—for shaping your message, crafting your delivery, shooting your video, and choosing your technology—are discussed in chapters 6, 7, 8, and 9.)

Web 2.0 Applications

Skype, TelePresence, YouTube, Facebook, LinkedIn, ooVoo, and Twitter are just some of the many Web 2.0 applications that have fundamentally transformed the way people communicate. One of the core elements of some of these applications is the ability to send and receive high-resolution and high-bandwidth digital video. Two of the novel video-based Web 2.0 applications introduced after

2003 are YouTube and Skype. Anyone with a digital camera that allows capture of low-resolution movies can now take a digital video and post it for the world to see on YouTube. Skype and its competitors are a class of software that revolutionizes telecommunications over the Internet. Skype is a computer-based application that facilitates video and voice communication through an Internet protocol. It connects computers with other computers for communication, and also can connect computers to telephones. Skype is the equivalent of an Internet-based videophone. Anyone can download the software and use it. With the addition of a webcam, any desktop or laptop computer can be converted to a videophone capable of communicating with anyone in the world who has a high-speed Internet connection and a webcam. For a fee, you can even use it to connect to regular telephones, using audio only.

Web 2.0 applications that use digital video are not just for consumers. New technologies are fundamentally changing the way businesses communicate. Telepresence is the name given to a new set of information technologies that provide a video and audio experience to individuals in which they feel as though they are in one location although they actually are in another. Cisco Systems' Cisco TelePresence is a leading-edge example of the technology. Introduced in 2006, Cisco TelePresence connects physically separated and distant conference rooms in ways that make you feel as though distant participants are sitting next to you. People all over the world can now meet virtually and "see" one another in the same room via high-definition images projected onto flat panel displays. A spatial audio system causes a person's speech to emanate from where the person is "sitting." The communication experience is like people sitting next to you instead of hundreds if not thousands of miles away. Many other companies—including Aethra, HP, LG-Nortel, Polycom, Radivision, Sony, Tandberg, and Vidyo—have developed similar systems.

Why is the idea of using video for communications so important? Obviously, such advanced video-based technologies offer dramatic cost reductions to organizations by reducing the costs of traveling to meetings as well as time lost in transit. Meetings no longer need to be constrained by your travel schedule, which generates numerous economies to firms.

But there is a more fundamental question: Why do many people prefer face-to-face conversations in the first place? Research on human conversation and communication provides a striking answer: Viewing one another during a conversation fundamentally transforms the ability for humans to understand each other. Many studies have established the importance of nonverbal cues in human communications. Albert Mehrabian, a psychologist who had a long and distinguished career at UCLA, derived the 7-38-55 percent hypothesis, according to which approximately 7 percent of communication is received through verbal transmission, 38 percent is received through vocal transmission (voice and tonality), and 55 percent through body language such as posture, gesture, and especially facial expression.[29] Although these percentages are likely not precisely valid for conversations you hold in your community, most people acknowledge the importance of body language and facial expressions to developing understanding. Perhaps it should be no wonder that e-mail communications and memoranda commonly fail to generate understanding and why holding a conversation over the phone to discuss complex issues can take a long time and yield unsatisfactory results.

ChangeCasting

Web 2.0 technologies hold the promise of fundamentally changing the way you can lead organizational change in your community. It is the combination of using web 2.0 technologies to implement specific processes that I refer to as web 2.1. ChangeCasting is a process that uses Web 2.0 technologies in a specific way to help you accelerate organizational change. Web 2.0 technologies can be used for the three principles needed for change based on commitment: conversing with your entire community simultaneously, protecting your community's conversation, and demonstrating that you listen to your community.

The ChangeCasting process comprises three steps that use Web 2.0 technologies (see figure 1).

Step 1 is creating web-delivered digital videos—ChangeCasts—for you to converse with your community.

Step 2 is providing web-enabled e-mail by which your community can respond anonymously to your ChangeCasts.

Step 3 is assessing anonymous feedback and demonstrating in your subsequent ChangeCasts that you listened and responded to your community's concerns.

Later chapters will build on this three-step process and identify specific guidelines to help you shape your message and deliver it using web-based video.

Step 1

Principle 1 is that you communicate to your entire community all at once. How can you accomplish this goal? Without the assistance of technology, it is difficult for you to communicate to everyone at once except through low-tech media such as e-mails and memoranda or through the chain of command, which is subject to filters and distortion. Alternatively, you could ask the community to stop work and take the time to gather for a general meeting, where you address them and they listen to your comments. This has high opportunity costs—the time could be better spent on something else—and is virtually impossible with a widely dispersed community. Now, with Web 2.0 technology, your community members can remain at their desks or work stations anywhere in the world and receive a video of you conversing with them. Practically simultaneously they can watch the video or wait until opportunity costs for doing so are not so great.

Using web-enabled video offers several benefits, the most obvious of which is cost reduction. In the past you could have made a series of videotapes or DVDs and shipped them around to different locations for people to watch. Doing so not only would have been costly and time consuming but also it required that each viewer or group have a video player. Now, the ubiquity of desktop and laptop computers makes it easy for anyone with access to a computer to view digital video online.

If you wanted community members to attend a meeting to hear you give a talk, they would need to interrupt their activities to do so. Now, they can watch your ChangeCast at a time that is convenient with their work schedule, for example, at the end of a shift or when they are not "fighting a fire." Even if community members don't view your message at precisely the same time, web-enabled video allows your community to watch it at nearly the same time; in fact, a larger number may end up watching it than the number that you could attract to a meeting. A web-based ChangeCast also offers the playback benefit: your community can watch the video again or return to it later to be sure that they heard your communication correctly. Meetings offer no such playback feature.

ChangeCasting, if executed properly (see later chapters), allows your community to see you up close and in person. You can communicate your emotions, facial expressions, and subtle body movements, all of which contribute to developing understanding but may not be observed during a speech or meeting and certainly are hidden in the case of e-mails, memoranda, and phone calls. Web-enabled video communications therefore allow you to communicate to everyone in your community all at once in a way that was not possible with older approaches and technologies.

Another benefit of using personal computers to distribute and view web-enabled video is that doing so can contribute to building trust. When people gather to listen to you, whether in the context of a formal speech or a meeting, the attendees subtly search the eyes of others in attendance for cues as to whether they should trust your words or not. If many community members fear losing something valuable, it may not take too many distrustful glances to negatively affect the entire group present at your talk. Change-Casts allow community members to view your communication individually and think about it before discussing it with others. Distribution of digital media can encourage community members to develop their opinion more independently than before, which can give you an advantage in building trust even when community members do compare notes.

Step 2

The second principle to enable change through commitment is to protect the conversation—to encourage people to come forward and share their ideas and thoughts with you in an open exchange. Indeed, if you do not protect the conversation, people will not speak their minds. And without candid feedback how can you understand which assumptions and conclusions to challenge? Changing your community members' minds is much easier if they can express their views without fear of criticism or retribution.

Information technology provides a new means for protecting your community's conversation. One of the most important factors that undermine conversations is fear of embarrassment, ridicule, and of being punished. All of us have been to meetings where it's quite clear that attendees may have reactions or useful observations to share, yet refuse to participate. They may say later that they were not asked for their opinion or were not listened to, even though they themselves chose not to speak up during the meeting.

One way to eliminate this fear and to invite higher levels of participation is to provide a way for comments to be made anonymously. Web-enabled e-mail can be made anonymous (discussed further in chapter 9). Anonymity means that you must commit to not knowing the origin of an e-mail and not trying to track down the sender, even if you do not appreciate its content. This promise, if kept, means that you are likely to receive some unpleasant e-mails. Yet it is probably better for you to hear critical comments directly from the source than to have such sentiments shared with other community members during coffee breaks and for you only to hear about it much later through rumor and innuendo. Hearing about members' concerns quickly allows you to respond in your conversation and snuff out rumors that can sidetrack and undermine your change effort. To be sure, reading unpleasant and even inappropriate comments requires a thick skin and requires you to frame a thoughtful response in your next ChangeCast. But receiving these messages and responding to them is a great way to build trust in your community.

Of course, anonymity means that people will not receive credit for their contributions, but it also means that they can't be embarrassed, ridiculed, or punished for their reactions. It may take awhile for your community to believe your promises of anonymity, but with time and consistency, your community will find your assurance of anonymity credible.

There is good news about the range of anonymous e-mails that you will likely receive. My conversations with ChangeCasting leaders suggest that just a small percentage of e-mails are unhelpful or inappropriate; the vast majority of responders attempt to engage the leader or offer congratulations for providing communication and being open to comments. Such communications signal a growth in trust and understanding.

Step 3

Demonstrating that you're listening also can be enhanced by web-based information technology. Reviewing the anonymous feedback that you receive via e-mails allows you to identify and respond to your community's most commonly reported issues and concerns. In your next ChangeCast, you can both tell and show your community that you listened. In fact, your ChangeCasts represent an opportunity to identify and educate your community on what appears to be the most pressing issues that have surfaced in feedback. This has an additional beneficial effect by disseminating to community members what others believe is important, which can move the community toward common positions and understanding, instead of extreme ones.

The old cliché that a picture is worth a thousand words is appropriate here and is consistent with the value of digital video for developing understanding. Suppose your community is concerned that you are not making promised investments in the organization. You can show them actual images of your investments such as construction sites, training and development activities, new IT infrastructure, and so forth. For those who don't perceive that new and desired behaviors are taking hold you can show visual evidence of people behaving differently, making heroes out of them for doing so. In a meeting, by contrast, it is difficult to respond on the spot with illustrations to demonstrate your investments.

Summary

The principles for enabling and accelerating change can be implemented in a new way in a Web 2.1 world. You can converse with your entire community more frequently using ChangeCasts. You can protect the conversation by providing web-enabled anonymous feedback, which allows you to receive comments that otherwise might not reach you. And, you will hear these comments faster and from a broader spectrum of people once your community believes that the comments indeed are anonymous. You can demonstrate that you understand your community's concerns, by simply restating it in your ChangeCasts and also by showing visual evidence of how you and the company have responded to their concerns. The ChangeCasting process holds the promise of enhancing your ability to lead and ultimately accelerate change.

6   *ChangeCasting Guidelines*
THE MESSAGE

KEY POINTS

⇒ Some guidelines for having a conversation *are* better than others.

⇒ Five guidelines can help you shape your message and improve your conversation in ways that can build trust, create understanding, and accelerate organizational change:

    —Be brief and regular (two-to-five-minute ChangeCasts every few weeks).

    —One idea for each ChangeCast.

    —Be real about today.

    —Aspire for tomorrow.

    —Formulate problems before solving.

Some conversations are better than others, and so, too, can some ChangeCasts be better than others. What can you do to ensure that your message invites conversation instead of silence? Chapter 6 presents five guidelines that can help you craft messages that can lead to a good conversation with your community. Although being a good conversationalist is something of an art form, experience and research have taught me that five guidelines are among the most important practices you can adopt to have a good conversation with your community.

1. Be brief and regular (communicate every week or two).
2. Focus on one idea per ChangeCast.
3. Be real about today.
4. Provide aspirations for tomorrow.
5. Be sure the community understands the problem before you offer solutions.

Message Guideline 1: Be Brief and Regular

The length of a ChangeCast is central to its effectiveness and depends very much on the expected attention span of your community. The length of the ChangeCast is intimately connected with the frequency of your messages, so length and frequency are considered one guideline.

How long can your community pay attention to and work to understand your message? Can they easily remember and recall it? How long should you speak in a ChangeCast? Fortunately, research on attention spans provides some useful guidance. Also, the experience of organizations that currently provide Web 2.0 video content is useful for determining the optimal length of your ChangeCasts.

Listening and understanding are hard work, especially because most people do not have long attention spans. Attention spans vary by age as well as by what activities someone engages in prior to listening. Two types of attention spans have been studied: focused attention span (unbroken attention with no lapse) and sustained attention span (the level of attention over an extended period of time that produces a consistent result). Numerous studies of continuous focused attention span have found it to be surprisingly short—seven to ten seconds.[30] After just a few seconds, minds need to drift for at least a moment.

When it comes to a sustained general attention span, some people can focus for longer, especially in a conversation, although the mind will still experience brief moments of drift. Studies indicate that the average general attention span is between ten and fifteen minutes for adults but as brief as two to three minutes for young children. The adult mind needs a sustained break lasting several minutes after focusing on something for fifteen minutes.[31]

Other studies indicate that adult attention spans may in fact be much shorter than fifteen minutes if the mind has been focusing on something else for a while. A study of college students found a surprising pattern across twelve lecturers. A. H. Johnstone and F. Percival observed students in over ninety lectures, across twelve different lecturers in chemistry, noting breaks in student attention. From their observations they concluded that a general pattern emerged. After three to five minutes of "settling down" at the beginning of class, a "lapse of attention usually occurred some 10 to 18 minutes later, and as the lecture proceeded the attention span became shorter and often fell to three or four minutes towards the end of a standard lecture."[32]

Since adults often are busy, engaged in work and other activities that might be mentally draining, you are probably safe in assuming an average attention span to be between three and five minutes.

Another way to think about attention span with respect to ChangeCasting is to examine what the market has determined. To explore this question, I analyzed all videos posted on the *New York Times* website, for their duration. A total of ninety videos were analyzed. The average video was 3.5 to 4.0 minutes long on all the days sampled. These statistics offer an appropriate benchmark because one can assume that a newspaper attempts to convey important information in a format that viewers can understand. In fact, during the period when the data were collected, the *New York Times* reportedly was experimenting with four- to seven-minute videos.

FIGURE 2. Video Duration at the *New York Times*

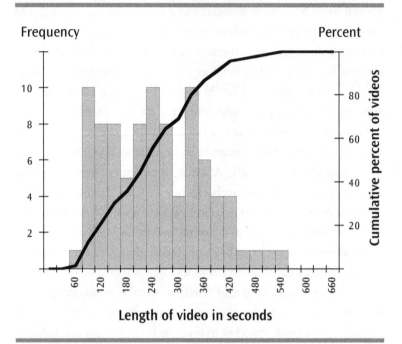

Length of video in seconds

The graph in figure 2 shows the ninety videos analyzed from the *New York Times* website in the form of a distribution curve. The horizontal axis is the length of video in thirty-second increments (durations are rounded to the nearest thirty-second interval). The vertical axis is the number of videos of particular lengths. The average video duration is 3.5 to 4 minutes. The shortest video was sixty seconds and longest was 9.0 minutes. Approximately 80 percent of the videos were shorter than 6.0 minutes.[33] The *New York Times* apparently has found that the optimal video length for capturing attention and conveying information is substantially less than the maximum average general attention span.

These data suggest that ChangeCasts should be fairly short. On the basis of my own experience with creating and using Change-

Casts and working with leaders who do so I suggest that a useful duration is two minutes and no longer than five minutes.

How frequently should you release ChangeCasts? The frequency of your ChangeCasts is related to their length. It is difficult to have a conversation if you are not communicating with your community with some regularity. For example, it is difficult to have a conversation if you are creating ChangeCasts more than a month apart. But your ChangeCasts should not be too frequent, either. You need to give your community time to view your message, think about it, and respond. Sending out ChangeCasts more frequently than once a week does not give your community enough time to reflect on your comments and respond to them.

In general, my experience has been that ChangeCasting every week or two—depending on the pace of change—is beneficial. One of our two CEOs ChangeCasts every two weeks and has maintained this frequency during as well as after rapid change. Once-a-month videocasts may be more appropriate if the change effort is not facing an urgent crisis. ChangeCasting at intervals longer than one month may send a signal that you are not really interested in sustaining a conversation with your community.

Message Guideline 2: One Idea per ChangeCast

Sometimes I find myself trying to communicate too much information and jamming too many ideas into a message, and this can be problematic. Communication can be defined as the exchange and progression of thoughts, feelings, and ideas with the goal of gaining a mutual understanding. Creating a shared understanding requires conveying an idea and then checking to make sure that the receiver understands it. In light of people's short attention spans and the need to create messages of short duration, it is important to focus and delimit the content of each ChangeCast so that its content can be grasped and verified. Conversation with your community is most effective when you present one idea at a time and then verify that the idea has been understood.

ChangeCasts of two to five minutes in duration are not long enough to present more than one idea. Talking about more than one idea may make the communication difficult to understand; it may also make it difficult for you to sort out the feedback and verify understanding from the anonymous e-mails you receive. Complicated or multiple ideas take more time to communicate and understand. Even communicating simple ideas may require somewhat longer messages. The number of ideas you can communicate in a ChangeCast is limited if you stick to the two-to-five-minute duration guideline.

Bundling several ideas in a single ChangeCast may result in your receiving too great a diversity of responses to be able to verify understanding. Furthermore, crafting a response to the diversity of responses in your next ChangeCast will be made complicated. If you are unable to verify understanding and provide a focused response because of a diversity of responses, you may end up destroying trust and undermining understanding, which amplifies uncertainty instead of reducing it. So the rule of thumb is to communicate one main theme in each ChangeCast, perhaps reiterating it several times instead of packing your messages with multiple ideas.

Message Guideline 3: Be Real about Today

Your community is smart and likely possesses much information from the trenches that you do not possess. They will trust you on the basis of their collective assessment of your character, goodwill, and ability and will use this knowledge to evaluate your statements. Providing an accurate and "real" assessment of your community's situation speaks of your character and ability. If they don't think that you are providing the real and complete picture of the current situation—one that is consistent with the information they have about the situation—your community may question your integrity. If you are not straight with them they may think that you don't trust them with the truth. If you don't trust them enough to convey to them the real situation, then why should they trust you? Concerns about the validity of the information you convey also may cause your community to question your ability because they may reason: if she doesn't know the real situation then how can she successfully lead us through the change? The absence of belief in your capability undermines trust. Whether you adopt ChangeCasting or some other approach for leading change, being real about today is important for maintaining and building trust within your community.

Message Guideline 4: Aspire for Tomorrow

Being real about today does not imply that you must be pessimistic about the future. In fact, you as the leader must have a belief and vision that the community will be in a much better situation in the future if it changes—your aspirations for tomorrow. With a commitment approach to leading change people need hope to help them overcome fear and drive change. Statements of aspiration should focus on the future but need not provide many specifics. For instance, a for-profit company need not specify that the company will be the leader in its industry in five years, particularly when it might presently be at the bottom of the pack of competitors. Such statements can be motivational, but they also can stimulate distrust if the community cannot imagine how the organization can make the necessary transition.

Your aspirations for the future need to be communicated in a way that reflects your character, goodwill, and ability. Statements of aspiration that are believable to your community will support all three. Statements that are not consistent with any one of these attributes can cause trust to unravel. To design statements that are believable it may be useful to start your change process with communicating aspirations that reflect your goodwill toward the community. Statements of this kind are very specific to each community and therefore you should try out your statements on people you trust before communicating them to the larger community. For instance, you may want to try out and test your statement of aspirations with selected community members, which will help you avoid the unraveling of trust. As your community begins to understand the direction of changes and uncertainty is reduced, more specifics can be added to your statement of aspirations.

Message Guideline 5: Formulate Problems before Solving

One of the most community-damaging biases of individuals and groups is jumping to conclusions or to a solution before ensuring that the right problem is being solved. Have you ever been in a meeting for the purpose of discussing a challenge, and before the problem is even explored, someone throws out a solution? What happens next? Often, either the solution is confronted by a room full of "devil's advocates" who tear apart the proposal, or someone else tosses in a rival solution. The ensuing conflict can quickly bring the meeting to a boil because the solutions are interpreted as being in the self-interest of the proposers, which ultimately leads to a politically based selection of a course of action to which not everyone is committed. All the while, the group fails to comprehensively formulate the problem. Individuals and groups who behave this way often end up solving the wrong problem and, to make matters worse, experience poor implementation. Eventually, the undefined and unresolved problem grows worse and leaders must backtrack to try again to define the real problem and find an effective solution. This pattern of "solving" the wrong problem is sometimes called "fighting fires."

Such outcomes occur because the proverbial invisible elephant is in the room and the group fails to see all sides of it at once. The inability to understand all the aspects of the elephant—the problem—naturally leads people to propose partial solutions that often disproportionately benefit them without first comprehensively formulating the problem. Once someone proposes a solution, the leadership challenge is amplified because the proposer's ego is triggered. When challenged, the proposer will naturally come up with reasons and logic to support his position and reject the challenge and often will expend great effort doing so. The proposer becomes defensive and resists new perspectives, and this triggers a wide variety of biases that undermine good decisionmaking and that oppose change.[34]

Allowing people to jump to the wrong conclusions can undermine your change efforts. Fortunately, you can shape your message and change it over time to help avoid this problem. Consider grouping your ChangeCasts in three phases. In the first phase, use your ChangeCasts to describe the challenges and opportunities your community faces. Be careful not to discuss solutions or implementation during this first phase of the conversation. When solutions are offered during this phase, ask people to reflect on the problem to make sure that it has been defined accurately and comprehensively. What problem does your community need to solve? Have you considered all aspects of the problem and its context? Your community will perceive good character, goodwill, and high ability—and hence is more likely to trust you—when you involve it in problem identification and consider the information individual members offer and listen to their opinions. By helping your entire community understand what the invisible elephant looks like you increase the likelihood of coming up with a comprehensive solution. You will also create common understanding in your community of the challenge that you face, which will encourage people to work together once a solution is decided upon.

In the second phase of the conversation, focus your Change-Casts on developing a solution to the commonly understood and comprehensively formulated problem. When most people understand the problem correctly, they can not only help brainstorm about solutions but also verify that the solutions proposed indeed address the entire problem. To be sure, some people are going to be eager to jump ahead to implementation once the problem is identified. It may be best to slow down implementation until, through ChangeCasts, your community understands the entire solution. Then your community will be in a better position to advance the implementation instead of questioning why some actions are being taken.

The third phase of the conversation is implementation of the solution. Imagine that everyone in your community understands the challenge as well as the solution and trusts the solution is a good one. Will this understanding and trust accelerate implementation? The central proposition of *Leading Change in a Web 2.1 World* is that if the organization understands the challenge, solution, and implementation plan and trusts you, in part because of the process you followed to arrive at the change, your goals and approach to change the organization are more likely to be their goals as well. If your community understands why the changes have to be made and why a particular implementation path has been selected, neither mistrust nor misunderstanding will delay implementation.

Summary

This chapter presents five guidelines for crafting your ChangeCasts. Following these guidelines can increase the likelihood that you will have a good conversation with your community, one that leads to accelerating organizational change. ChangeCasts should be brief, and appear every week or two. Because of their brevity and the need to verify the ideas that are communicated, each ChangeCast should have only one idea. Your messages should be realistic about both the challenges faced today and your aspirations for the future. It is vital that neither you nor your community latch onto a solution before the problem has been comprehensively formulated. Following these guidelines can help you build your community's trust, help you solve the right problem, and speed efforts to find a solution and implement it.

7

ChangeCasting
Guidelines
THE DELIVERY

KEY POINTS

➡ Be yourself (don't act or speechify in front of the camera).
➡ Be compassionate (show that you understand your community's perspective).
➡ Be direct (make your communications neither evasive nor blunt).
➡ Invite conversation (ask specific questions that need a response).
➡ Use symbolism (images advance understanding).

No one likes to be embarrassed by looking ridiculous or nonsensical to others. Broadcasting web-enabled video communication to an entire community creates a situation in which many leaders are particularly concerned about the potential for embarrassment. In fact, just as some members of your community fear change, some leaders may fear making fools of themselves and this fear may stop them from using ChangeCasting.

To help you muster the courage to give ChangeCasting a try, this chapter gives you the information you need to create effective videos that make you look good. There are five guidelines for enhancing the delivery of your message: be yourself, be compassionate, be direct, constantly invite conversation, and use symbolism. These delivery guidelines can help you build trust and create understanding to help you accelerate organizational change.

These five pieces of advice for improving your delivery may sound like common sense. They are. Nonetheless, many leaders fail to fully appreciate their importance and to use these guidelines when conversing with others. Since web-enabled videos can be made quickly and at low cost, it is easy to check your video to verify that you have used the guidelines or to redo your ChangeCast if necessary. By employing these delivery guidelines you can not only reduce fears about how you come across in ChangeCasts, but also stimulate within your community positive perceptions of your character, goodwill, and ability. Positive perceptions of these three factors can be relied on to increase your community's trust in you. The guidelines, if followed, also will advance your community's understanding of your change effort.

Delivery Guideline 1: Be Yourself

Can you remember the first time you were videoed? If you are like most people you may have become self-conscious and displayed shyness, nervousness, or goofiness. Even after much experience, some people find it very hard to relax in front of the camera. Their self-consciousness affects how they act on camera, which impacts how others watching the resulting video perceive their persona. The difference between how people are perceived on and off camera is vital to leading change.

A common word associated with being yourself is "authenticity." Authenticity is defined as being trustworthy and genuine—worthy of belief. Being yourself in front of the camera signals to your community that they can believe in you. If you behave differently in front of the camera than you do in person, your community members might conclude that you are trying to pull the wool over their eyes and that you are not believable.

Your behavior on camera lets your community know that you are a real person, which enhances their perception of your beneficence and goodwill because they are more likely to identify with you. If authenticity is an important trait to display during your ChangeCasts, what can you do to ensure that you are authentic and are perceived as such?

First, do not use a script. Most people adopt a more formal tone, cadence, and way of speaking when they read from a script. Your community immediately will recognize when you are reading from a script, which will invite questions and suspicion. Why does she need a script? What is he trying to avoid saying? Using a script raises doubts in your community that will undermine your authenticity.

If you focus on communicating one idea in each ChangeCast, as recommended in the prior chapter, you won't need a script. If the idea you want to communicate can't be remembered off the top of your head then your message probably is too complex. Simplify it.

A second aspect of communicating authentically is to speak naturally, conversationally. Don't change your speaking style for the video. For instance, if you normally use your hands when you speak, then use them. If you naturally pause when speaking, then take pauses. If you typically prefer speaking while standing up, then stand up during your video. If you occasionally turn your head when speaking to help formulate your thoughts, then go ahead and do so. Speak in a way that is comfortable to you.

Try to imagine that you are speaking to another person instead of to a camera. Actors train to do this; it is something that can be learned, but it may be difficult for you and me to do so and do it well without practice. An effective trick is to put a graphic of someone's face next to or over the camera lens and focus on that instead of the camera lens. This trick may seem silly, but in fact, the image will help to fool your brain into adopting a more conversational style of communication. Alternatively, you can speak to someone standing behind the camera. If these tricks don't work for you, experiment and develop your own tricks to help you be authentic in your videos.

Many people are afraid of making flubs in videos, but you needn't be too worried about this. Your objective is not to compete with a professional news anchor. Your goal is to demonstrate your authenticity, and an occasional flub lets people know that you are a real person and are not reading from a script. So if you trip over a word or have to correct yourself during the video, don't worry about it and just keep going.

One of the great benefits of recording digital video, such as from a computer-based webcam, is that capturing images is easy and costs little except your time. The ease of recording and its low cost mean that you can delete a video and quickly start over if you don't like it. When I first began developing and using ChangeCasts, I would close my office door and record my messages using a webcam. Early on, I tried to remember a speech or read from notes on my screen. The results were not very good—I definitely could see that I was not being authentic and did not look authentic.

I must admit that occasionally I would have a brain cramp during a recording and forget what I wanted to say in midsentence. When this happened I would laugh about it, delete the video, and try again. I also would view my video and look for any part of my communication that did not seem authentic. To be sure that you are not changing your natural communication patterns, ask those with whom you have had face-to-face conversations to preview your ChangeCast. I would on occasion ask my colleagues to take a look at my video to give me their reaction. If you are uncomfortable with asking others to preview your ChangeCasts or if you want a more experienced review, you will find described in the last chapter a service where you can e-mail your ChangeCasts to be previewed and evaluated.

With a little practice and feedback, you quickly will find your stride in creating authentic ChangeCasts. Even with little or no preparation, you will be able to create ChangeCasts in one or two takes. After my first few months of ChangeCasting, I frequently was able to create a video in just one take. Occasionally I choose to do a second take, but additional takes are rarely necessary. If they are needed it may be a signal that you are not clear enough on the message of your ChangeCast and that you should clarify your message before recording it.

Delivery Guideline 2: Be Compassionate

You may wonder why "Be compassionate" is a guideline separate from "Be yourself." You also may wonder whether you have to be a passionate speaker to convey an emotion such as compassion. In fact, being a compassionate or empathetic speaker does not equate to being a passionate or emotional speaker. In the context of ChangeCasting, you show compassion when you demonstrate genuine concern for your community. You demonstrate empathy when you show that you understand others' feelings. You don't need to demonstrate strong emotions passionately in order to communicate compassion and empathy. You can do so simply by recognizing how others in your community are affected by the situation you collectively find yourselves in.

Being compassionate is important because it builds trustworthiness—through beneficence and goodwill—in your community's eyes. Compassion and empathy positively contribute to how others perceive your character and goodwill. Remember that character and goodwill are two of the three core elements underpinning trustworthiness. Therefore, being compassionate can directly contribute to how your community views your trustworthiness and can contribute to building trust.

One useful approach for demonstrating compassion is to provide a narrative of someone in your community who has faced his own challenges with either the old way or new way of doing things. People generally are interested in stories about members of their community. So, early on in your organizational change process you might consider describing how the old way of doing things was frustrating for a worker or for customers, which signifies at least one reason for changing the organization. Later in your change effort you might chronicle the difficulties faced by a worker in trying to adopt new behaviors and how the worker overcame adversity. Such stories indicate to your community that you understand and are compassionate about their situation.

Delivery Guideline 3: Be Direct

What does it mean to be direct with your communications? Being direct does not mean being blunt, crude, stark, or devoid of the subtlety and qualification needed to talk about complex situations. Being direct means not dancing around or evading difficult topics. Instead, you should help your community understand the full complexities of the challenges to be faced.

Being direct reminds me of a piece of advice my father gave me many years ago: be a straight shooter, which means get to the heart of the matter quickly and don't try to manipulate or sugarcoat your communications. He also told me that being a straight shooter requires heeding a second piece of advice: my father urged me to be thoughtful about the language I use to communicate with others. The language you use matters to a surprising degree.

It is vital to use terminology and language with which your community is familiar and comfortable. Using sophisticated and erudite language and jargon will alienate your community and cause them to call into question your ability as well as your compassion. For instance, academics like to use the word "heterogeneity" when the word "diversity" often will do. Relying on overly simplistic language leads to a similar outcome because it does not demonstrate respect for your community's intelligence. Being direct and using language your community understands and is comfortable with will likely enhance perceptions of your trustworthiness.

Delivery Guideline 4: Invite Conversation

As a young professor at the Olin Business School I would ask during my classes, "Does anyone have any questions?" Most of the time no hands would go up and no one would ask a question. In time I figured out that I was asking the wrong question.

My request for questions was generic and nonspecific. It was easy for students to avoid raising their hands because I did not require them to do so. Eventually, I learned to ask specific questions that required an answer, such as "Name one thing you will take away from today's class"; "What is the one thing I can do to improve today's class?"; or "What is one thing I should not change about today's class?" These questions are specific enough so when I posed them to students they couldn't avoid answering the question. A response was required. Also, they had something specific to focus on. After a few days of asking specific questions in class, the pump had been primed and the questions flowed as if a high-pressure spigot had been opened wide.

The same is true for using ChangeCasting to generate dialogue. A general request for questions and reactions will elicit little response. The more specific your questions, the more likely people will feel compelled to answer them. Open the spigot for anonymous feedback by asking specific questions that demand a response. For instance, try asking your community to send you feedback on one thing they liked about your most recent ChangeCast as well as one thing they didn't like about it. In general, encourage your community to communicate with you by sending you queries or comments anonymously. Keep doing so in each and every Change-Cast and responses eventually will flow.

Delivery Guideline 5: Use Symbolism

Communication is accomplished through more than just words. Indeed, the effectiveness of ChangeCasting is predicated on the communication benefits of your community seeing your body language and facial expressions. Symbols, imagery, and narratives provide another important avenue of communication. For instance, an appropriate metaphor or analogy can go a long way in helping your community understand your message. Equally important is showing images that demonstrate investments you are making to implement your strategy, display workers engaging in the new and desired behaviors, and provide indicators of success, such as customer testimonials.

My favorite kind of symbolism involves making heroes and heroines out of those in your community who support the change effort. For instance, in one of our client companies, information technology was an important element of the strategy being implemented, but most employees did not understand the importance of IT. The CEO filmed a ChangeCast from his firm's central computer room, which allowed him to show that investments in IT were large and ongoing. He also highlighted recent successes within the IT group in implementing change efforts. His ChangeCast made heroes out of the IT group and helped the entire community understand the importance of IT in the new strategy.

Summary

Chapter 7 presents five delivery guidelines that can help you overcome your fear of embarrassment concerning ChangeCasts and help you build trust and create understanding in your community for your change effort. Be yourself, be compassionate, and be direct; together these can affect how your community views your trustworthiness. Invite conversation by asking specific questions and use symbolism to focus your message. Together these guidelines can contribute to advancing your community's understanding of and commitment to your change effort.

8

ChangeCasting Guidelines
THE VIDEO

KEY POINTS

➡ Speak to the camera eye-to-eye.

➡ Be close to the camera.
—Keep your eyes level with a point one-third from the top of the screen.
—Keep the bottom of the frame above your chest.

➡ Be distant from the background.
—Keep a distance of 10 feet between you and the background.
—Avoid backgrounds with bright colors.

➡ Dress for digital images.
—Don't wear pinstripes or small repeating patterns.
—Don't wear red or yellow clothing.

➡ Choose the right audio and light.
—Avoid harsh shadows, and bright light that washes out the video.
—Use a separate microphone only if there is background noise or you are distant from the camera.

Not only your message and its delivery contribute to building trust and creating understanding; so does your appearance on video. You needn't be a professional videographer to make yourself look good on video; you just need to follow a few basic video techniques to make it easier for people to pay attention to and understand your messages. Chapter 8 offers five video guidelines to improve your ChangeCasts:

1. Speak to the camera eye-to-eye.
2. Be close to the camera.
3. Be distant from the background.
4. Dress for digital images.
5. Use the right sound and light.

Video Guideline 1: Speak to the Camera Eye-to-Eye

A variety of different camera angles can be used to film a speaker. Your position with respect to the angle and location of the camera—the framing of the shot—is important because it can affect how your audience perceives you. The camera can be positioned relatively close to or far away from the speaker. It also can be positioned at an angle so that the speaker is looking off to one or the other side of the camera or straight at it. Figures 3 and 4 display several of these different ways of framing a speaker

As an illustration of how camera angles can matter, panel A in figure 3 shows a popular way for framing a speaker that can be seen on many television-based news shows: the interviewee is at eye-level but off to one side of the video frame looking at an interviewer, who is off camera on the other side. The interviewee never looks directly at the camera during the interview but directs her gaze to the interviewer. Though interesting and artistic, the framing is not conducive for creating a conversation between you and your viewers. Viewers of such interviews perceive that they are outside the conversation, watching it like a fly on the wall, instead of perceiving themselves as part of the conversation. With this camera angle and framing, viewers don't feel like that they are part of the conversation and can lose attention and focus.

FIGURE 3. Framing Your Video

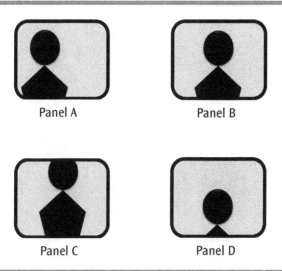

Panel A   Panel B

Panel C   Panel D

Panels C and D display other camera frames that can be distracting to the viewer. In panel C, the camera frame cuts off the speaker's head. If the camera is located below your eye level and the angle is from below, looking up at you, it conveys a sense of large size or power. Panel D shows the opposite: the frame cuts off the speaker's entire torso and there is a lot of space above his head. Or the camera may be looking down on you, which suggests a diminutive stature. Neither of these images invites viewers to imagine that they are having a conversation with you.

For ChangeCasting it is imperative that the camera lens be at eye-level with you, and that you are centered in the video frame, looking directly into the eye of the camera, as shown in panel B. Speaking directly to the eye of the camera creates the impression in viewers of your ChangeCasts that you are looking at and speaking directly to them. They will feel like they are part of the conversation and will be more attentive to and focused on what you are saying.

FIGURE 4. Distance to the Camera

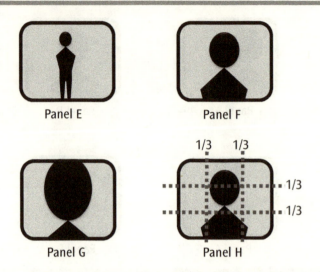

Panel E   Panel F

Panel G   Panel H

Video Guideline 2: Be Close to the Camera

Another important dimension of framing is shot length: the camera's distance from you and the background. Framing refers to the extent that your image fills the video frame. For you to maximize the benefits from video-based communications, your community must be able to see your facial expressions and body language, and this requires close-up framing.

Figure 4 shows various frames according to the speaker's distance from the camera. Panel E shows a long or medium shot, where the speaker is seen at a distance. If you are distant from the camera while recording your ChangeCasts, it will be difficult for viewers to see your facial expressions and body language. If, on the other hand, you are shot in extreme close-up (panel G), your image will appear to invade viewers' space. Don't be a space invader because it puts off the viewer. Furthermore, an extreme close-up may show a few too many details of your facial features! The ideal shot distance should allow viewers to feel as though they are having a face-to-face conversation with you. You should frame the video so that the bottom of the video screen is just above your chest, as is shown in panel F. This distance is easily gauged: imagine superimposing a tic-tac-toe pattern on your video screen, as in panel H. In fact, most new digital cameras come with such a feature for helping photographers frame their shots. You should frame your image so that your eyes line up with the top horizontal line of the tic-tac-toe design, about a third of the way from the top of the screen. Panels F and H show how you should be framed by the video.

Video Guideline 3: Be Distant from the Background

It is only natural for viewers to take in an entire video image, which includes not only at you but also the background. The background can be distracting, especially if there are lots of colorful items or if there is movement. The best way to avoid background distractions is to make sure the background is distant from you and the camera. It is best for you to be at least 10 feet from any objects in the background when you are captured on video. This distance will help keep the camera as well as the viewer focused on you and what you are saying and viewers will not be distracted by something superfluous.

Video Guideline 4: Dress for Digital Images

You need to think carefully about what you wear in front of a digital camera to avoid unwanted visual effects caused by patterns or colors in your clothing. One of the most common of these patterns is a moiré pattern; you have probably on occasion observed moiré patterns in your own digital photographs and pictures as well as on your computer. The patterns arise because digital images are made up of thousands of pixels, small minuscule rectangular blocks of color, and pixels can interact to create unwanted visual effects in the video. Moiré patterns can be distracting and cause viewers to focus their attention on the pattern instead of on the content of your video.

Avoid moiré patterns by wisely choosing your wardrobe for ChangeCasts. Do not wear pinstriped patterns, houndstooth patterns, or garments that have small repeating patterns in contrasting colors. Wear garments with solid colors or large-repeating patterns to help ensure that the focus remains on your conversation instead of your clothes.

You should also be aware that certain intense colors, especially when they cover large portions of a video screen, can be distracting and fatiguing to the human eye. Bright yellow and red are the most fatiguing, so avoid these bright colors in both your garments and your background.

Video Guideline 5: Getting the Right Sound Quality and Light

Video technology, which combines sound and visual recording, has advanced by leaps and bounds over the past two decades. Today you can purchase digital video recording devices at prices ranging from $25 to $5,000 with remarkable capabilities to adjust sound and light settings to deal with very varied conditions (equipment options are discussed in the next chapter). So your chances of achieving good sound and visual quality are very high, and there is not too much to worry about. Nonetheless, following a few simple rules will help ensure that your video looks good—not too dark, high contrast, or washed out—and sounds good.

The best light quality is a strong diffused light that doesn't cast shadows. High-powered lighting such as that used by professional videographers is not needed to create good video images. When filming indoors, be certain not to have powerful overhead lights casting shadows down across your face. If you do see shadows on your face, moving a few feet one way or the other in relation to your light source should eliminate them. Do not allow bright light sources to be filmed. A bright light source behind you also can be a problem because it can cause your image to be washed out, or possibly create a halo effect, where the bright light causes the image of your face to be dark and the outline of your head blurry.

Most webcams, camcorders, and video recorders can give you acceptable sound quality with their built-in microphone, as long as you are positioned close enough to the microphone or are using a telescoping microphone. In a noisy environment or if you need to stand quite far from your video recording device, you may need to use a lavaliere (also called a lapel) microphone, a small mike attached to your clothing (such devices are discussed further in chapter 9).

Summary

How you look on camera and are framed by the camera are important. Your goal is to keep your audience focused on you and keep them from being distracted. In this chapter I introduced five basic video guidelines to make it easier for people to pay attention to and understand your messages: speak to the camera eye-to-eye, be positioned close to the camera, be positioned distant from the background, dress for digital images, and avoid pitfalls with lighting and sound.

9

Technology for Managing ChangeCasting

Leading Change in a Web 2.1 World is predicated on using information technology to manage a conversation with your community. It is therefore indispensable to describe some of the technologies that you can use to make ChangeCasts available to your entire community and to receive anonymous feedback from it. Fortunately, the current state of information technology is such that low-cost, easy-to-use devices and software make it simple for practically anyone to start ChangeCasting with little effort. Technology is not a barrier to ChangeCasting, especially for leaders of small communities. But large organizations offer an opportunity for leaders to use more sophisticated technology that can provide greater

FIGURE 5. IT for ChangeCasting

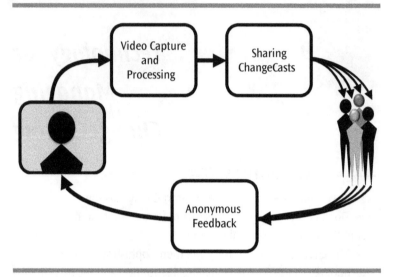

functionality. This chapter provides an overview to help you under-stand what technologies are appropriate for your community.

We all know that information technology advances at an extraordinarily rapid pace, and new devices and new software appear all the time. This rapid technological change makes it virtually impossible to provide comprehensive and current infor-mation about the different technologies available. What this chapter can achieve is an explanation of and brief introduction to technology categories. This overview will not make you an expert on video-related information technology, but it will introduce you to factors that will likely affect your technology decisions. For ongoing updates on technology advances introduced to the market and a more comprehensive listing of technology alternatives, please visit the ChangeCasting website (www.ChangeCasting.com).

Figure 5, "IT for ChangeCasting," shows the three types of information technology required for ChangeCasting:

1. Video recording (also called "capture") and processing of footage
2. Sharing ChangeCasts with your community
3. Providing a means for your community to anonymously send you feedback

You have many options for each one of these activities. To help simplify the range of options, we discuss low-cost, moderate-cost, and high-cost alternatives. As you might expect, as the cost increases so too does functionality—the number of bells and whistles—which will enable you to do more with your ChangeCasts. But, as you will discover, doing more with your ChangeCasts is not always desirable.

Recording and Processing ChangeCasts

Recording ChangeCasts can be surprisingly easy. Perhaps the simplest technology is a webcam, a video recording device connected to a computer or computer network, often using a USB port, so that you can see the video on your monitor screen or send the images to other computers. A large number of manufacturers sell webcams for between $25 and $125—the higher the price, the better the video, light balancing, and sound quality. Many computers now have built-in webcams as part of their standard hardware, as well as the software needed to record audio and video. As an example, Apple computers come equipped with iMovie, which enables video and audio capture. All of these options are low-cost and provide adequate and in some cases excellent quality video for ChangeCasting. For instance, I ChangeCast to an organization of a dozen people that I manage from hundreds of miles away. I make my ChangeCast videos while sitting in front of my computer, using a ninety-dollar webcam.

Low cost and ease of use are great advantages of webcams, but you will get better video quality if you invest in a moderately priced camcorder, an electronic device that combines a video camera and a video recorder in one unit. Costing between $100 and $1,000, camcorders give you many options for creating higher-quality video. Camcorders can have superior lighting control, come with their own boom microphones to improve audio quality and reduce noise, and have a jack to connect a lavaliere or other external microphone. A microphone will add an additional $25 to $500 to the cost of your moderate video recording system, but it can greatly enhance sound quality.

In addition to enabling better video and sound quality, camcorders are portable, so you can conveniently record your ChangeCast in ever-changing locations. For instance, you may want to record your messages in different locations where you can visually show capital investments or introduce community members who are making a difference.

High-priced video cameras, ranging in price from $1,000 to $5,000, also are an option. They cost more than moderate-priced camcorders because they come with bells and whistles such as image stabilization, professional-quality lenses, choice of aspect ratio, film-like progressive scanning, HDMI (high-definition multimedia interface) output, and a built-in microphone that supports different audio formats, as well as other features that can produce movie-quality results. A downside of moderate- and high-priced video cameras, other than price, is that recordings must be transferred to your computer, which adds an extra processing step, although most new video cameras, especially digital ones, have rendered the transfer step relatively easy by making it possible to connect to the computer via a USB cable.

My general recommendation is that a webcam offers a great way to start ChangeCasting. Webcams are inexpensive and easy to use, and video can be shot and processed quickly. Webcams are perfect for office settings. If you plan to make ChangeCasts from different locations, then you may find a camcorder or video camera a worthwhile investment.

High-priced video cameras provide many bells and whistles that facilitate editing, but I don't actually recommend that you use them. Editing your video can undermine your authenticity. Still, some editing features are useful. For instance, you may find value in inserting an introductory slide with your name, date, topic, or organization's logo on it. Or if some of your community members speak a different language from you, you may want to transcribe your conversation so that it appears as subtitles at the bottom of the video. In that way your community can see you, hear you, and read what you are saying, which can help to build trust and create understanding. Another potential role for editing is to reinforce a message by adding text during or at the end of your ChangeCast, for example, listing your community's core values at the end of your video.

The low-cost option for processing your video is simple—the delete key on your computer keyboard! If you don't like your ChangeCast, delete it and start over. In general, editing is not needed to make outstanding ChangeCasts because authenticity is the vital element. Remember that it is okay not to look like a television news anchor in your ChangeCast. You just need to be you.

For those of you who want to add simple edits such as inserting a title slide, you will find that many basic editing tools are available at low cost and may be free or bundled with your computer's software. For instance, iMovie, bundled with MacBook, is an easy-to-use, if limited, video editor. Most providers of video editing software offer beginner packages for under $50. These moderate-cost programs likely will provide all that you will need to insert slides and add captions.

High-end video editing software is available, of course, and these packages can cost well over $1,000. Such video editing software offers powerful tools to design eye-catching professional layouts. However, this advanced editing takes time and may be counterproductive: in many ways it can undermine authenticity. Consequently, I do not recommend allocating much time and money to advanced editing software and techniques.

Component 2: Sharing ChangeCasts with Your Community

As with video capture and processing, a wide variety of options exist for sharing ChangeCasts with your community. The range of options varies with cost, security, and time needed to manage the videos. Although I recommend the low-cost options for video capture and processing, for sharing the message with your community I recommend moderate-cost options for smaller organizations and high-cost options for large organizations.

When I first began ChangeCasting I would simply e-mail videos to my community members. It was low-cost and easy for me to capture the video on my webcam and e-mail the ChangeCast directly. It took me about five to ten minutes every few weeks to prepare and send a ChangeCast. E-mail also was convenient for my community because with a single click they could be watching the video on their computer in seconds. What could be easier?

The low-cost approach works well when you have a small community, but it can quickly run into three challenges. First, ChangeCast videos can be large files that consume network resources and hard-disk space. A three-minute ChangeCast could use from three to two hundred megabytes of memory, depending on how the video was captured and processed. Because of size and security concerns, some organizations limit the file size of e-mail attachments, which can create distribution problems.

Second, with e-mailed files you are relying on your community members to manage storage and retrieval of the files on their local computer, as some may want to return to a ChangeCast at a later date. Managing files does require a minimum level of computer system awareness and sophistication, which not everyone in your community may have.

Third, e-mail makes it easy to forward your ChangeCasts—but you may want to make this difficult, especially if they might be forwarded outside your organization. You may want to include community-specific and sensitive issues in your ChangeCasts to which you would not want individuals outside your community to have access. Many organizations already have policies that limit e-mailing company-specific information outside their organization, which should mitigate at least some security concerns. Nonetheless, people do make mistakes when it comes to forwarding e-mails, and security issues should not be taken lightly.

So how will you make your ChangeCasts available? One option is to simply e-mail them or post them on YouTube or another similar public site that allows you to create password access; then share the password with your community. As mentioned earlier, e-mailing videos is problematic because of e-mail and computer space limitations. Posting ChangeCasts on a public site can raise security concerns.

A more secure, moderate-cost alternative is to create a webpage on an intranet—a private computer network that uses Internet protocol technologies to securely share any part of an organization's information or network operating system within that organization—and post files to your webpage. You can then send out a link via e-mail to let your community know that your next ChangeCast is available. Using a website resolves several of the issues that arise with e-mailed or publicly stored videos. For instance, this technology does not require community members to be sophisticated about storage and retrieval and managing their computer system; the organization's information technology group can manage security. Thus a secure intranet can help you better manage security issues.

Such a secure internal website offers a good moderate-cost option but does have a few weaknesses. Once it is set up, additional effort is required to maintain it, which involves keeping the webpage up-to-date, posting videos, and sending out notices that new ChangeCasts are available. You also might choose to manage who is to receive them, provide multilevel security, manage when the video should be taken down, and so forth. Managing and securing intranet web pages can be difficult and time-consuming because they are not specifically programmed for these tasks. If webpage-management issues such as page maintenance and security are important, which is likely to be the case in large organizations, you may want to adopt a high-cost but also a high-functionality solution like a content management system.

High-cost alternatives typically involve specialized work environments or video content management systems. These systems also can manage anonymous feedback, which is discussed in the next section. For instance, a company called Qumu offers sophisticated systems that manage and deliver enterprise video. Their video content management system makes it easy to post video, manage who is to receive it, send out notices that new ChangeCasts are available, provide multilevel security, manage when the video should be taken down, and so on. Cisco also offers enterprise video content management systems that can be used for managing ChangeCasts.

Collaboration software is another technology that can provide a secure environment within which to host ChangeCasts. Microsoft's SharePoint can be programmed to provide video management and security, although it may not be as valuable as systems designed specifically for video content management. All of these systems are expensive to purchase and must be set up and run by information technology professionals, but they do provide additional functionality, especially with respect to security. These sophisticated systems are most appropriate for large businesses with multiple locations that are or plan to be operating in numerous countries.

Component 3: Anonymous Feedback

Perhaps the most difficult—but also the most vital—aspect of ChangeCasting is achieving the third information technology activity: enabling anonymous feedback. Arranging for anonymous feedback is not technically difficult, but convincing your community that feedback is anonymous is extraordinarily difficult. The difficulty arises because your community will likely fear that the electronic communication is not anonymous. Indeed, television shows and movies have contributed to the belief that practically all digital communication can be traced to its author—and in many instances the belief is justified.

Fortunately, your community's skepticism can be overcome, but doing so requires two conditions. First, you need to adopt an information technology system that can provide anonymous e-mails. This signals to your community that indeed you want feedback to be anonymous. In addition, you must put your personal reputation on the line by publicly stating that no matter what technology is used, you will ensure that the system is anonymous. If this promise is violated, your trustworthiness will take the hit. Over time, as you demonstrate your commitment to this policy, your community's fear will eventually subside and your conversation will meet with greater participation and take on more meaning.

Receiving anonymous feedback is easier than you think. Several low-cost (even some free) tools are available for you to seek anonymous feedback. A number of companies provide online survey tools such as Survey Monkey, Zoomerang, Instant Survey, and others that allow you to quickly construct a survey and enable responses for each ChangeCast that you make. You can include a link to the survey with e-mails that deliver or announce your next Change-Cast. These companies provide third-party credibility that you are indeed serious about providing anonymous feedback. One caveat is that some of these companies put "cookies" on a respondent's computer for their internal tracking purposes, which naturally stokes the fear that identities could be discovered if someone got around the legal and firewall protections provided by these firms.

Low-cost survey tools typically have more advanced features that are available for a moderate cost. For instance, Survey Monkey is free to use for a small number of questions and to a limited number of survey participants. More advanced features and high usage limits typically can be had for as little as two hundred dollars a year. Of course, information technology professionals in your own organization can probably design and program equivalent systems, for a price. But your community members will likely know that the system was produced in-house, which may raise suspicions that the feedback is not truly anonymous.

As stated, enterprise video management systems as well as collaboration software typically can be configured to enable anonymous e-mails for feedback. These systems provide an integrated environment for both distributing ChangeCasts and receiving anonymous feedback, but your community may not trust that the systems really are anonymous if your information technology department runs them. It is especially important that you continuously reinforce to your community your commitment to anonymity. If anonymity is violated even once you will lose a fundamental building block of your trustworthiness and will invite fear and misunderstanding back into the conversation.

Summary

ChangeCasts require three information technology activities: capturing video and processing it, sharing ChangeCasts with your community, and providing a means for your community to anonymously send you feedback. Low-, moderate-, and high-cost alternatives are available for each activity. Inexpensive webcams work well in most settings, although camcorders are useful for making videos in different locations. Editing your video isn't recommended except merely to add an introductory title or subtitles, for the convenience of non-native speakers of English. The type of software needed for distributing your ChangeCasts and receiving anonymous feedback depends on the size and complexity of your organization.

10 *Did ChangeCasting Improve Performance?*

By now you may be eager to learn how Gen and William fared in their tests of leadership. Did our CEOs' attempts to lead change in the context of a Web 2.1 world benefit their communities? Were Gen and William able to sail through rough seas and lead their organizations to higher levels of performance or was the storm too great a challenge for them to successfully navigate? Did both firms perform well? Did one firm do better than the other, and if so, which one? What role did ChangeCasting play in these outcomes?

Understanding how ChangeCasting was adopted in their organizations and how their communities responded to it offers at least one way to evaluate its use. Another more concrete way to evaluate the effects of ChangeCasting is to assess the financial performance of Gen's and William's organizations after they adopted ChangeCasting. The firms' performance, as measured by relative stock prices, is reported before, during, and after the adoption of ChangeCasting. Obviously, many factors of which ChangeCasting is just one could explain changes in relative firm performance. In this chapter I attempt to evaluate the effects of ChangeCasting in the context of other alternative explanations for changes in each firm's performance.

The Challenge of Adopting ChangeCasting

Faced with a pressing need to lead change across their organizations, both Gen and William agreed to become ChangeCasting leaders. Getting them to adopt ChangeCasting took some convincing, as both executives initially were reluctant to adopt a new methodology when their companies were at a crossroads. At the time, neither CEO had any direct experience of ChangeCasting nor could they rely on the experience of other organizations because the process and guidelines were new; they would be among the first to try it. Neither organization had an infrastructure in place specifically for ChangeCasting. ChangeCasting was a new paradigm for community members as well as for leaders, which presented its own set of issues: the community would have to come to an understanding of why its leaders would send out brief messages and solicit anonymous feedback. All of these issues were impediments that would have to be overcome if ChangeCasting's process and guidelines were to be successfully adopted.

Yet both Gen and William intuitively understood that good communication was important for their organizations and that communication would take on an even more vital role in light of the changes they were committed to leading. Both CEOs acknowledged that leading change would involve taking risks and that initiating an ongoing conversation with their community would give them an additional tool for managing these risks. Both leaders also had a support staff that could help them implement the process and guidelines of ChangeCasting. In Gen's case, she had a technical marketing organization that could help with shooting and processing video as well as processing feedback. In William's case, he had a corporate communications team that was used to shape messages for communicating to the workforce. Each took a courageous leap to try out a new process and agreed to become ChangeCasting leaders.

Gen's Experience

Gen launched her first ChangeCast in late fall. (To maintain Gen's and William's anonymity, specific dates and years are not identified.) Thereafter she made ChangeCasts about every two to three weeks. The length of her videos ranged from one to five minutes. Eventually Gen concluded that two to two and half minutes was the best length. Her initial videos discussed her company's strategic challenges as well as elements of the reorganization that she had recently launched. Each ChangeCast focused on one primary topic, although sometimes more than one topic was discussed. Some of the early ChangeCasts also introduced to her community vital activities that they knew little about. In these ChangeCasts, Gen made heroes out of community members, like those working in the information technology group, by explaining the importance of the activity to the company's future along with changes taking place. Because of shifting business patterns, information technology was taking on a more strategic role in the company—a role that many in the community did not initially understand or appreciate.

Many of the ChangeCasting topics were based on feedback from Gen's community. In making ChangeCasts, Gen would ask her community to e-mail reactions, questions, and comments in response to the video. To ensure that all community members from around the world would understand her comments, the marketing team added a text bar on the bottom of the video, like the ones that accompany TV broadcasts, so that viewers could read along if they wanted to. Gen received roughly two hundred to four hundred e-mail responses after each ChangeCast. After a few weeks of reading the e-mails that flooded in response to her ChangeCasts, Gen asked her marketing group to sift through the responses received from each ChangeCast and identify the two or three most frequent reactions, questions, and comments. Typically, these common responses provided the topic of the next ChangeCast. Within a few months of launching the ChangeCasting process, Gen wanted to encourage more to join the conversation so she had an icon installed on the desktop of every personal computer in the company so that with a single mouse click everyone with access to a computer could e-mail a response to her ChangeCast.

Gen followed the three main principles of ChangeCasting: She frequently sent out brief ChangeCasts. She enabled feedback. And she demonstrated that she listened to her community by using the feedback from the last ChangeCast to shape the next part of the conversation. One principle that was not initially adopted had to do with anonymity of feedback. In the beginning, all e-mail messages were sent directly to Gen via standard e-mail, which identified the sender. Gen was surprised and pleased that some felt comfortable enough to offer critical comments even when their identities were known. Comments reached Gen anonymously after messages started being processed through the marketing department and changes were made to the e-mail system. But even though some in the community were willing to provide critical comments, I suspect it took some time for Gen's community to become convinced that they had nothing to fear—a precondition for a candid conversation—by responding candidly to her ChangeCasts.

Gen also generally adhered to the ChangeCasting guidelines concerned with crafting her message, its delivery, and her videos: Her messages tended to be around two minutes long. In them she was real about today yet aspirational about the future of her firm. Most ChangeCasts focused on one theme and she attempted to discuss challenges and gain agreement about them before trying to solve them.

Gen's delivery also lived up to ChangeCasting's prescribed guidelines. Gen did wonderfully at being herself in the videos. She told me later that it would take her no more that fifteen minutes to create a ChangeCast. She would show up where the marketing group had set up the camera, which often was at different locations so that she could symbolically show where she was making investments to grow her organization. Sometimes she would not find out the primary concern arising from the last ChangeCast until she arrived at the location; yet she responded to it on the spot, without additional preparation. In fact, she may well have come across the most authentic when she learned about the primary concern of her community and addressed it on the spot. This spontaneity enabled her to be direct and compassionate. It was difficult for Gen not to be authentic in such situations. Then, to conclude each ChangeCast, Gen would be certain to invite reactions, questions, and comments, which consistently signaled her openness to conversation. Her video would be completed in one or two takes and was not further edited except for the addition of text at the bottom of the screen and a title page. For Gen, the key to being herself was not to prepare too much.

Her marketing staff did well adhering to the recommended guidelines for recording video. Gen almost always was positioned close to the camera with her eyes located about one third of the way down from the top of the screen and she spoke directly to the lens of the camera. Whenever possible, she was positioned distant from any backgrounds. In some early ChangeCasts, she wore clothing such as striped shirts that was visually distracting, but eventually she dressed to avoid such distractions.

She chose appropriate information technology for the size and global nature of her firm. She used a digital video recorder instead of a webcam. Adding a title to each ChangeCast was a nice touch that made the video more professional looking. A special internally secure website was set up to host the videos, and a link to each new video was e-mailed to the community members. WorldCo already had a policy in place prohibiting leaking internal electronic information, so over the many years that Gen has been ChangeCasting, none have ended up on YouTube or any other external Internet site.

All in all, Gen adhered to ChangeCasting's process and guidelines. Gen continues to ChangeCast and has been doing so for many years. One big change in her ChangeCasts compared to when she started is that she now invites other leaders to respond to questions from her community, thereby further expanding the conversation and helping others to become ChangeCasting leaders.

William's Experience

William's adoption of ChangeCasting's process and guidelines was very different. He began ChangeCasting in the late spring. Unfortunately, neither he nor his staff adhered to ChangeCasting's three principal process steps or to its guidelines. His first ChangeCast offers a telling illustration of the ways in which his adoption ignored the prescriptions described in this book.

Whereas Gen used her existing marketing personnel to take the video, add text to it, and post it, William's communications team hired a professional videographer who staged the video, shot it, and edited it, which was expensive—several thousand dollars. The edited video was over ten minutes long. It was shot in a formal setting with plants obviously staged to make the video appear professional. William was framed to the side of the shot, and rather than looking directly into the camera, he was instructed to look off to the other side, as if he were having a conversation with an interviewer who was not in the frame. This guideline is rather common for television talk shows and news exposés but does not generate the trust necessary for ChangeCasting conversations. Indeed, William's eyes never met the lens of the video recorder during the entire interview. The video obviously had been edited because there were several breaks in William's monologue in which his image faded out and then back in again.

William's communication was sincere. He was truthful about the firm's situation and hopeful about the future. Nonetheless, his message suffered from four weaknesses. First, perhaps because of the way it was edited or the coaching he received from the videographer, he did not appear at ease or authentic. His remarks seemed prepared, which can undermine the building of trust.

Second, William talked about many issues, which is why his initial ChangeCast lasted more than ten minutes. Perhaps because of the cost, he tried to pack a wide variety of issues into a single video. It is difficult for viewers to remember, let alone process, a conversation that broaches multiple issues. Even though I as a viewer was a very interested party, my attention began to fall off after the first five minutes of the video.

Third, instead of focusing on the central challenge and discussing solutions later, William began laying out solutions and actions. It was too easy for viewers to conclude that decisions had already been made and that William was looking for buy-in and acceptance, not for his community to be in the conversation.

Finally, William did not invite feedback. Not only did he fail to ask for comments, reactions, and questions but no system was set up to enable his community to anonymously engage in a conversation with him. The company was in difficult straits, and fear was rife. Absent an explicit invitation to communicate anonymously, the assumed order of the day for community members was, and remained, keep your head down and accept the one-way message.

William had a few other executives make web-enabled videos over the next couple of months. But those displayed similar characteristics as his initial effort, in the message, its delivery, and the video. No feedback mechanism, let alone an anonymous one, was put into place. Messages were one directional and did not stimulate a conversation between William and his community.

The Communities' Responses

A few months after Gen started ChangeCasting she missed creating a new video on the date expected. She described walking down the hall the next day and being accosted several times by community members inquiring as to why she hadn't posted a new ChangeCast. Before Gen could respond, they pleaded with her not to stop providing the videos and allowing feedback. Her colleagues enjoyed seeing Gen and hearing her thoughts directly. This type of response is not at all uncommon among the several large organizations that have adopted ChangeCasting.

Over the next couple of years I checked periodically with midlevel managers on how ChangeCasts were received. In general, they had become part of the fabric of the organization and were well-received because it made what the CEO was saying and also how she was saying it visible to the community. Put differently, ChangeCasting helped Gen build trust and create understanding that she believed contributed to accelerating organizational change.

With time, I think it is fair to say that Gen's community engaged in a conversation with her and became much more aware of the challenges the firm faced. Such awareness did not occur overnight. Indeed, as will be seen, it took well over a year for Gen's ship to begin to turn and improve its performance. But, the ship did turn and many in her community will acknowledge that ChangeCasting played at least some role in helping Gen lead change and score highly on her fundamental test of leadership.

In contrast, William's ChangeCasting efforts had little impact on his firm. After a few months of sporadic attempts, William stopped ChangeCasting. Speaking with several midlevel managers, I found that William's videos had not built trust nor created understanding. In fact, the videos may have reinforced the impression that top management did not want to listen, did not want to engage in a conversation. In an unfortunate turn, web-enabled video may not have benefited the firm and may even have contributed to generating additional fear and resistance to change.

The Two Companies' Financial Performance

These assessments provide a context for interpreting the actual financial performance of each firm, as measured by the price of its shares on the stock market. To characterize stock market performance in a relative and comparative way, weekly share prices were collected for each company over multiple years. The two companies' weekly stock price series were modified to make them both more readily comparable to each other and to rule out a variety of factors that might provide alternative explanations for each company's performance.

First, I wanted to eliminate marketwide factors that may have caused fluctuations in each firm's stock price. To do this I adjusted the companies' stock prices with respect to a broad market indicator, the S&P 500, by dividing the weekly stock price series by the corresponding value of the S&P 500 Index. The resulting price series represented changes in stock prices not due to changes in the general economy. For instance, if a company's stock price increased 10 percent but so too did the S&P 500, then the adjusted price series for the company would remain constant.

Second, in order to evaluate whether company performance improved or declined over time in response to adopting Change-Casting, both ratios were normalized to 1 roughly a year before ChangeCasting was implemented. The normalization made it easier to compare the share prices directly. If the stock outperformed the S&P 500, the adjusted and normalized price series would rise above 1. If the stock price underperformed the S&P 500, the adjusted and normalized price series would drop below 1. If the stock price moved with the S&P 500 then the ratio would remain 1. Thus, the plotted ratios describe how each firm performed relative to the S&P 500, a broad market indicator.

Figure 6 shows the adjusted and normalized price series for the two companies' share price performance over several years (the actual dates and duration of time series are not disclosed to retain the confidentiality of our two firms). Circle A indicates when Gen implemented ChangeCasting. Before she implemented Change-Casting, her firm's adjusted and normalized stock price remained roughly around 1, which means her stock price performed no differently than the S&P 500.

The adoption of ChangeCasting had no immediate impact on the adjusted and normalized price of Gen's company; for more than a year after the adoption of ChangeCasting it stayed about the same. About a year and a half later, however, Gen's adjusted and normalized price strikingly pulled ahead of the S&P 500, rising above 1. The share price's climb began to accelerate and then continued to climb steadily above the performance of the S&P 500. By the end of the reported price series, Gen's company had outperformed the S&P 500 by climbing to a ratio of about 1.75, which equates to rising 75 percent more than any changes experienced by the S&P 500. Put differently, for every 10 percent increase in the S&P 500 Index over the time period, WorldCo stock price increased 17.5 percent.

The performance of William's company, indicated by the gray line, began similarly to Gen's firm during the reported adjusted and normalized price series. At first the share price moved with the S&P 500, which is indicated by the fact that the adjusted and normalized price series remained close to 1. Before he adopted ChangeCasting, William's firm experienced a precipitous drop in share price in relation to the S&P 500—market expectations for his firm declined dramatically. Perhaps it was in response to this decline that William agreed to try something new like Change-Casting, which was adopted during the time period indicated by circle B.

Shortly after William provided web-based videos to his community, the share price increased in relation to the S&P 500, but this proved fleeting; the price eventually continued to decline. By the end of the price series, displayed in figure 6, the value of MandACo's share price had plummeted and the company was fighting for survival. (The temporary improvement may have had more to do with a move to dispose of some assets and a commitment to make certain investments than with the adoption of web-enabled videos.)

The increase of stock price of Gen's firm compared to the S&P 500 is suggestive of Gen's having navigated her leadership challenge, whereas the performance William's firm is indicative that his firm was unable to successfully weather the storm. Less clear is the role that ChangeCasting played in contributing to Gen's success and William's failure. Did ChangeCasting play a vital role in Gen's success? Could William have saved his firm if he had executed ChangeCasting properly?

The two price series suggest that ChangeCasting can play a vital role in leading change, even though, from a statistical point of view, the two firms are a small sample. Many factors not easily measurable or observable by me or other analysts might explain

FIGURE 6. How Did Gen's and William's Firms Perform?

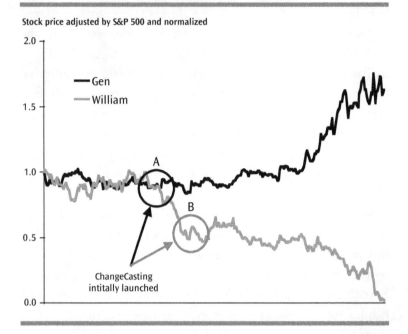

Stock price adjusted by S&P 500 and normalized

why Gen succeeded in changing WorldCo and William was unable to improve MandACo's performance. Perhaps Gen's community had a deep and well-functioning leadership team that responded to her leadership initiative and William did not have an equivalent well-functioning team. Perhaps Gen enjoyed the goodwill of her community and William did not. Gen's company, or its market position, may have had some hidden "good" quality that allowed it to change and grow whereas William's company or market position did not possess this quality. If so, it might be this good quality that explains both the successful change in performance and why Gen adopted ChangeCasting. A statistician would also point out that with a sample of two, it is difficult to make any clear statistical inference about the impact on firm performance of adopting ChangeCasting.

Statistics offer little help in drawing conclusions from two case studies, so it is up to you to draw your own conclusion. My hope is that the narrative setting forth the context of Gen's and William's situations and their stories along with the performance data will help you come to your own conclusions about ChangeCasting's contribution to leading change.

11

Should You Adopt ChangeCasting?

The stories of Gen and William provide two real-case studies that explore the use and benefits of ChangeCasting for the CEOs of two substantial and complex organizations. Gen implemented the ChangeCasting process and its guidelines, found it sufficiently valuable to continue using it, and led a successful change effort that produced a substantial improvement in performance over a time frame of several years. William, in contrast, tried using web-enabled video but without implementing either the ChangeCasting process or its guidelines. His change efforts were not ultimately successful. The narratives of these two CEOs' experiences and outcomes are suggestive that ChangeCasting might be a valuable tool to help lead and accelerate change.

Of course, not every leader is a CEO, so one important question is whether ChangeCasting creates value for you in your position. Does ChangeCasting offer you a new tool to advance your capability for leading change? To help you explore these questions, I will share with you a few thoughts on the conditions when Change-Casting is likely, or unlikely, to be useful. I also discuss some of the main challenges you might face in adopting ChangeCasting. But please don't worry—they are relatively easy to overcome. Then, if you choose to adopt ChangeCasting, I offer a few tips on the next steps for becoming a ChangeCasting leader and ways to get additional help if you need it.

When Should You Use ChangeCasting?

ChangeCasting offers value to leaders who find themselves in one or more of four types of situations.

A Geographically or Temporally Dispersed Community

ChangeCasting may be particularly valuable in situations where it is costly or difficult to communicate with all the members of your community simultaneously, for example, if your community is dispersed across multiple locations. Gen's firm has many locations on multiple continents. William's firm has scores of locations across North America. The geographic dispersion of locations makes it extraordinarily costly and difficult for any leader to speak to his or her entire community all at once.

Even if all workers are based at the same location, it nonetheless may be difficult to schedule community-wide meetings. For instance, consider how hard it is to find a time to meet with all the members of an outbound sales force. Or, consider a hospital or a police force, which have round-the-clock shifts and a substantial group of the community members must remain on duty at all times. It is extraordinarily difficult for a leader in these environments to engage the entire community in a conversation because rarely can the entire community be together. Multiple shifts in the same geographic location and difficulties in scheduling community-wide meetings present a type of dispersion—a temporal dispersion— that makes it costly, if not impossible, for a leader to converse with the entire community simultaneously. ChangeCasting may prove to be valuable whenever there is geographic or temporal dispersion of

community members because it overcomes the conversation challenges created by dispersion.

Advantages for Large and Small Settings

ChangeCasting may prove valuable in large organizations—
roughly, those with more than a few hundred members—whether
they are for-profit, government, or nonprofits. Obviously, the larger
a community, the more costly and difficult it is for leaders to gather
their community to get a conversation going. Furthermore, having
a meaningful two-way conversation becomes increasingly difficult
as the size of a community increases, because meetings with a large
number of participants create fear that inhibits conversation and
undermines any possibility of a meaningful, protected conversation. ChangeCasting may offer the only feasible method for
holding a conversation with a larger community.

ChangeCasting may also be valuable in smaller organizational
units found within or across a larger community, for instance
where change needs to take place within a business unit, division,
group, or team. To lead change in these organizational units, conversations need to take place within a subset of the entire community. ChangeCasting can be useful to leaders at all levels of the
organization to stimulate a conversation within their own organizational unit. Video communication and anonymous feedback can
stimulate a conversation that might otherwise not occur during a
team meeting because your community members may fear sharing
their comments with others. Thus, ChangeCasting may prove valuable to team leaders, group leaders, division leaders, and business
unit leaders as well as CEOs.

Conversations across Multi-Organizational Communities

Another challenge for leading change arises when a need for coor-
dination arises across multiple units of an organization in which a

leader may not possess formal authority in all units. Consider the case where a leader in one division needs to lead change that affects other divisions and suppliers. How can the leader engage a multitude of organizational units in the absence of formal authority? ChangeCasting offers at least one way to launch a conversation that can span several organizational units. However, doing so may require the advanced information technology and content management systems discussed in chapter 9.

Leading change in government agencies may pose an even greater challenge to spanning organizational barriers. National defense, counterterrorism, economic growth, and education, as well as many other activities of governments, require the cooperation of multiple agencies or subunits within and across agencies. A recent pressing example is the challenge of integrating and synthesizing information gathered across the U.S. intelligence community, which comprises eighteen separate units and offices:

Director of National Intelligence
Under Secretary of Defense for Intelligence
Air Force Intelligence
Army Intelligence
Central Intelligence Agency
Coast Guard Intelligence
Defense Intelligence Agency
Department of Energy
Department of Homeland Security
Department of State
Department of the Treasury
Drug Enforcement Administration
Federal Bureau of Investigation
Marine Corps Intelligence
National Geospatial-Intelligence Agency
National Reconnaissance Office

National Security Agency
Navy Intelligence

To combat terrorism this community must not only innovate but also continuously adapt to the changing strategies of a variety of terrorist organizations. It would be a very good candidate for ChangeCasting (assuming the availability of secure communication by means of the advanced information technology tools highlighted in chapter 9). Thus, ChangeCasting offers a new tool for launching a conversation to enable change across broad, loosely or closely affiliated but separate communities, which otherwise might be extraordinarily costly and difficult.

In some ways, nonprofits face dilemmas similar to governmental organizations. In domains ranging from health care and social services to universities and nonprofit networks, it is often the case that many independent or quasi-independent entities need to collaborate and innovate to remain current. Yet the relative independence of nonprofits and units within them makes leading change costly and difficult if not all but impossible. For instance, in a confederation of hospitals with which I am familiar, leading change to share best practices and install systems that capture economies of scale is extremely difficult because there is no central authority that can command the change. How can someone lead change effectively across the confederation?

Each major city in the United States typically has an amalgam of social services that are specialized to provide various support services. Almost anyone with aging parents in need of various support services is aware of the difficulty in arranging for a continuum of care outside a hospital setting because they have to stitch together the various social services provided by different organizations. Rare is the city where someone has successfully led efforts to successfully coordinate social services to eliminate the gaps that occur between specialized services. How then can someone lead change effectively across multiple social services organizations?

Units within a university often are fiercely independent and jealously protect their turf. With separate identities and functions and an undertone of competition, how can leaders from the central administration or from individual academic units successfully lead change across university departments?

Some nonprofits use a type of network organizational structure such as that used by the Public Broadcasting Service. PBS has a centralized leadership organization in Washington, D.C., that generates programs and provides an IT infrastructure, but it also has a large number of independent stations that, typically, are not compelled to adopt any specific programming or policies. In order to remain relevant and adapt to the rapid changes brought about by the Internet and related information technology and by shifts in funding, how can individuals effectively lead change across such networks where participation and the adoption of policies and practices are largely voluntary?

In all of these multi-organizational communities, in the absence of a centralized authority for leading change (see chapter 4), gaining commitment through conversation is the main option for leading change. ChangeCasting is designed explicitly for the purpose of launching and sustaining such conversations and offers a new methodology that may greatly enhance the ability of leaders in these multi-organizational and networked organizations to pass their tests of leadership.

Leader Has Time Constraints

ChangeCasting may be valuable for small communities when the leader faces time constraints for engaging the community in conversations. Leaders who travel a great deal to bring in business, manage multiple constituencies, or lead virtual corporations may have little time for the one-on-one conversations needed to build and maintain commitment. Such busy leaders can ChangeCast from anywhere so long as they have a laptop computer with a webcam and a way to distribute their ChangeCasts. Also, anonymous responses from community members can be read anywhere and anytime, which will enable conversations that otherwise might not take place. In essence, ChangeCasting can increase the velocity of conversations for small communities.

When Not to Use ChangeCasting

ChangeCasting offers a new methodology that is useful to leaders at every level of an organization, but it is not for everyone. In chapter 4 I highlighted two broad approaches to leading change, compliance and commitment. Both of them have their place and usefulness, but ChangeCasting offers little value for leaders who rely on the compliance approach to leading change. The principal value of ChangeCasting is that it enables leaders to build trust and create understanding to generate commitment. If the challenge a leader is facing calls for a compliance approach or the organization's culture is one of compliance, ChangeCasting is unlikely to offer much value. Indeed, William's firm had a history and culture of using a compliance approach for managing, which may have played a pivotal role in shaping how William's communications personnel implemented web-enabled videos.

Summary

ChangeCasting may be valuable for every leader, whether a CEO or a team leader, who wants to adopt a commitment approach to leading change and finds it costly or difficult to hold a conversation with her community. Conversations are likely to be costly and difficult wherever community members are geographically or temporally dispersed or when change efforts span multiple communities. Conversations also are costly and difficult in small communities if the leader does not have much time for one-on-one conversations. If your situation is characterized by any of these conditions, I urge you to consider becoming a ChangeCasting leader.

Challenges in Adopting ChangeCasting

If you are considering becoming a ChangeCasting leader, it is important for you to become aware of a few challenges that you may encounter. I have observed several organizations' attempts to implement ChangeCasting, and three challenges sometimes emerge. Fortunately, all of them can be overcome without too much effort.

The greatest challenge to becoming a ChangeCasting leader is overcoming your own fear of embarrassment. For some people, talking to a video recorder is easy—they don't get embarrassed easily and come across very naturally and authentically on video. But such people are the exception. Many others are worried about how they look or sound on video and shy away from ChangeCasting. Or, if they do proceed, they overprepare or use a script, which undermines authenticity. If you are anxious about recording yourself on video, I recommend shutting your office door and using your webcam to record your own video without anyone present. That is what I do. Record a video for about two minutes; then stop and watch it. Don't stop recording if you stumble over words or forget what you meant to say—just keep recording. When you are done, if you don't like what you see and hear, hit the delete key and try again. No one will be the wiser. You may be surprised at how quickly you can get over fear of embarrassment when you have full control of your own recording and can record in isolation without others watching. My own experience suggests that with a little practice you will create a ChangeCast in which you come across as authentic in one or two takes.

The second challenge is to ChangeCast on a regular basis. It is important do so—to overcome inertia and other claims on one's time to make creating regular ChangeCasts a priority. As you know, leading a community often results in a busy and hectic schedule. If you don't commit to a routine for creating your videos—say, every two weeks or so—it can be surprisingly easy to put it off until later. Letting your schedule slip is problematic in two ways. Delaying your ChangeCast slows down the conversation with your community, delaying the building of trust and the creation of understanding, which are vital to generating commitment. Also, if you establish your ChangeCasting frequency and then deviate from it, you may find it harder for your community to trust you. Conversing with your community on a regular basis is vital to building trust. If you announce one schedule and do another, your community will pick up on your inconsistent behavior and will not place their trust in you. Even if you did not commit to a schedule for your ChangeCasts, your community will infer an expected frequency from your first half dozen or so videos. Deviation from the expected frequency may be viewed as a sign of a lack of commitment on your part, which will hamper your change efforts. However you look at it, you will slow your change efforts if you don't ChangeCast on a regular basis.

The third challenge is that your community may not believe that the feedback you request through e-mails is anonymous. Most will think that you can track down the source of an e-mail if you really want to. A few technical fixes for this problem are offered in chapter 9, but the best solution is anchored not in technology, but in your character. If you say that the e-mails are anonymous, then you must keep them anonymous. Even if an e-mail uses harsh or inappropriate language, it is important not to undermine your integrity by tracking down the source of an unflattering e-mail. Keeping your word impacts your community's perception of your trustworthiness. All it takes is one slip of seeking out the messenger and "shooting," and your integrity will be undermined for quite a while.

All three challenges—fear of embarrassment, irregularity of ChangeCasts, and lack of anonymity—can undermine the conversation that you hope to have with your community. But each can be overcome with a little attention, effort, and commitment, and doing so will advance your ability to lead change.

Next Steps for Becoming a ChangeCasting Leader

I hope this book has gotten you excited about the possibility of becoming a ChangeCasting leader. The information technology revolution that made the need for leading change more frequent and more difficult also has produced tools that can be used to build trust, create understanding, and accelerate organizational change.

The key to getting started is to try it out! Chapter 9 offers several low-cost ways to begin ChangeCasting. You may just find that it is more fun than you imagined. If you do realize value from ChangeCasting then make the investments needed to support it in your community.

Trying anything new often is difficult. To help you become a ChangeCasting leader and for additional resources, technical information, and, if you wish, to arrange for evaluations of your ChangeCast, I encourage you to visit www.ChangeCasting.com.

If you try ChangeCasting and gain experience with it, I invite you share your experiences with others at the ChangeCasting website to help build the community of leaders who have found it useful.

NOTES

1. Wikipedia, "Web 2.0" entry (http://en.wikipedia.org/wiki/Web_2.0). The Wikipedia article contains a rich description of the full range of technologies and web applications associated with the term.

2. See M. J. Wheatley and M. Kellner-Rogers, "Bringing Life to Organizational Change," *Journal for Strategic Performance Measurement*, April–May, 1998; D. Miller, "Successful Change Leaders: What Makes Them? What Do They Do That Is Different?" *Journal of Change Management* 2, no. 4 (2002): 360; A. Raps, "Implementing Strategy," *Strategic Finance* 85, no. 12 (2004): 49. See also the website "Manage: The Executive Fast Track: Organization and Change: Methods, Model, and Theories," to learn of the variety of theories and processes available for leading change (www.12manage.com/i_co.html).

3. Numerous studies on leading change have been undertaken in business and the social sciences, and a large variety of prescriptive approaches are available. Some notable writers on leading change are Michael Beer, *High Commitment, High Performance* (San Francisco: Jossey-Bass, 2009); Warren Bennis, *On Becoming a Leader* (Boston: Basic Books, 2009); Daniel Goleman, with Richard E. Boyatzis and Annie McKee, *Primal Leadership* (Boston: Harvard Business School Press, 2002); Lawrence Hrebiniak, *Making Strategy Work: Leading Effective Execution and Change* (Upper Saddle River, N.J.: Wharton School Publishing, 2005); John P. Kotter, *Leading Change* (Boston: Harvard Business School Press, 1996); David Nadler, *Champions of Change* (San Francisco: Jossey-Bass, 1998); Charles O'Reilly

and Michael Tushman, *Winning through Innovation: A Practical Guide to Leading Organization Change and Renewal* (Boston: Harvard Business School Press, 1997); James O'Toole, *Leading Change: The Argument for Values-Based Leadership* (New York: Ballantine Books, 1995); and Bob Quinn, *Deep Change: Discovering the Leader Within* (San Francisco: Jossey-Bass, 1996).

4. Porter's *Competitive Strategy: Techniques for Analyzing Industries and Competitors* (New York: Free Press, 1980) sharpened awareness of the dangers to a company of attempting to adopt two positions that were in economic conflict—for example, starting up new product lines to differentiate an organization's offering to customers and also trying to achieve the lowest costs compared to competitors. The company could get "caught in the middle" and succeed at neither strategy because the strategies involve inherent economic trade-offs.

5. For a comprehensive academic treatment on the history of and approaches to organizational change, see W. Warner Burke, *Organization Change: Theory and Practice*, 2nd ed. (Thousand Oaks, Calif.: Sage, 2008). See note 3 for more sources on leading change.

6. The definition of the term "failure" may vary across studies. In most studies, it refers to the failure to achieve expectations. It also can mean the reversal of acquisitions and reorganization efforts.

7. This economic definition is sometimes referred to as Knightian uncertainty, after the early-twentieth-century economist Frank Knight, the author of a seminal work, *Risk, Uncertainty, and Profit* (Boston: Hart, Schaffner & Marx, 1921).

8. Daniel Kahneman and Amos Tversky coined the term "prospect theory" for the idea that people tend to fear the loss of something valuable more than they care about receiving future rewards. The theory, developed in a study of decisionmaking under conditions of risk, is that people strongly prefer avoiding losses to capturing gains (see Kahneman and Tversky, "Prospect Theory: An Analysis of Decisions under Risk," *Econometrica* 47 [1979]: 313–27). Kahneman won the 2002 Sveriges Riksbank Prize in Economic Sciences in Memory of Alfred Nobel, also known as the Nobel Memorial Prize in Economics, for his insights. For a comprehensive discussion of prospect theory, see http://en.wikipedia.org/wiki/Prospect_theory.

9. The fear of loss manifests itself in many meaningful and powerful ways. Jealousy represents the fear of loss and is known to be a very powerful emotion that causes people to take extreme actions, including murder.

NOTES 139

10. I thank Kurt Dirks for recommending the terminology of compliance versus commitment, which is referred to by many authors. See, for example, Michael Beer, *High Commitment, High Performance: How to Build a Resilient Organization for Sustained Advantage* (San Francisco: Jossey-Bass, 2009).

11. Jay Luvaas, "Military History: Is It Still Practicable?" *Parameters,* Summer 1995, p. 86.

12. Understanding of the issues inherent in the compliance approach gained much from a subfield in economics: the principal-agent problem, or agency dilemma. This literature explores the problem of motivating one party, an agent, to act on the behalf of another, the principal. So long as the agent's activities are monitored at low cost, an authority approach where the principal has the authority to direct the agent can work well. Of course, the principal has to possess complete knowledge about what the agent should do for his or her authority to be effective. When monitoring is costly, however, the literature provides insight into the design of pay-for-performance employment contracts that financially either reward or punish workers depending on the outcomes achieved by the workers along with how precisely the outcomes can be measured. The fundamental trade-off is between allocating risk and insurance for productive work. For more information, see http://en.wikipedia.org/wiki/Principal_agent_problem.

13. For a detailed account of this episode, see David Halberstam, *The Reckoning* (New York: William Morrow, 1986), Chapters 4 and 5.

14. This account of the five-dollar-a-day wage structure and Ford's sociology department is drawn from Michigan Department of Natural Resources and Environment, "The Assembly Line and the $5 Day—Background Reading" (www.michigan.gov/hal/0,1607,7-160-17451_18670_18793-53441-,00.html). See also Halberstam, *Reckoning*, chapter 5.

15. Halberstam, *Reckoning*, Chapter 5.

16. Footnote 12 introduces the economic field of principal-agent problems. One particular theory within the field addresses what is known as the "multi-task principal-agent problem." Many jobs are designed to include multiple tasks, which may differ in their cost of monitoring by a principal. In such cases, economic theory predicts that workers will focus their energies on tasks that are measured, or cost little to measure, to the detriment of tasks that are not measured, or are costly to measure. A variety of economic mechanisms can be used to mitigate multitask problems. For a more detailed treatment see Bengt Homston and Paul Milgrom, "Multitask

Principal-Agent Analyses: Incentives, Contracts, Asset Ownership, and Job Design," *Journal of Law Economics and Organization* 7 (1991): 24–52.

17. For a summary see K. Dirks, R. Lewicki, and A. Zaheer, "Repairing Relationships within and between Organizations: Building a Conceptual Foundation," *Academy of Management Review* 34, no. 1 (2009): 401–22.

18. Theories of communication tend to posit that there are three listening modes. Competitive or combative listening is when people are more interested in promoting their own point of view than in listening to what others have to say. Passive or attentive listening is when an individual listens without interruption, which implies that understanding is assumed instead of needing to be verified. Active or reflective listening is when a feedback process is required to verify that the listener understands the communication. It is the last mode that is most costly for individuals to engage in because it requires great amounts of attention, effort, and time. For an illustration of a professional curriculum that emphasizes the three listening modes,   see: http://professionalpractice.asme.org/MgmtLeadership/Management/Three_ Basic_Listening_Modes.cfm.

19. Aristotle, *Rhetoric*, can be found online at http://www2.iastate.edu/ ~honeyl/Rhetoric/.

20. For a seminal research article that summarizes current thinking on trust within organizations see R. C. Mayer, J. H. Davis, and F. D. Schoorman, "An integration Model of Organizational Trust," *Academy of Management Review* 20, no. 3 (1995): 709–34.

21. See, for example, D. L. Ferrin, K. T. Dirks, and P. P. Shah, "Direct and Indirect Effects of Third Party Relationships on Interpersonal Trust," *Journal of Applied Psychology* 91, no. 4 (2006): 870–33.

22. See A. D. Galinsky and others, "Power and Perspectives Not Taken," *Psychological Science* 17, no. 12 (2006): 1068–74. Four experiments and a correlation study explored the relationship between power and perspective taking. In experiment 1, participants primed with high power were more likely than those primed with low power to draw an E on their forehead in a self-oriented direction, demonstrating less of an inclination to spontaneously adopt another person's visual perspective. In experiments 2a and 2b, high-power participants were less likely than low-power participants to take into account that other people did not possess their privileged knowledge, a result suggesting that power leads individuals to anchor too heavily on their own vantage point and insufficiently adjust to others' perspectives. In experiment 3, high-power participants were less accurate than control

participants in determining other people's expressions of emotion. These results suggest that power is an impediment to experiencing empathy. An additional study found a negative relationship between individual difference measures of power and perspective taking. Across these studies, power was associated with a reduced tendency to comprehend how other people see, think, and feel.

23. Robert L. Shook, in his book *Turnaround: The New Ford Motor Company* (New York: Prentice-Hall Direct, 1990), provides a fascinating account of the decline and resurgence of the Ford Motor Company in the late twentieth century.

24. See, for example, Rafael Aguayo, *Dr. Deming: The American Who Taught the Japanese about Quality* (New York: Simon & Schuster/Fireside, 1991).

25. Shook, *Turnaround*, p. 86.

26. Ibid.

27. Ibid.

28. See Shook, *Turnaround*, for a comprehensive account of the changes Ford undertook in creating its new culture.

29. See Albert Mehrabian, *Silent Messages: Implicit Communication of Emotions and Attitudes* (Belmont: Wadsworth, 1971). Mehrabian's research, published in 1971, was derived from experiments that investigated the communication of feeling and attitudes measured as factors that contributed to "liking." Different content in communications therefore may lead to different ratios. Nonetheless, the vital insight is that nonverbal cues can be very important to developing understanding.

30. D. Dukette and D. Cornish, *The Essential 20: Twenty Components of an Excellent Health Care Team* (Pittsburgh: RoseDog Books, 2009), pp. 72–73, state that a focused attention span without any lapse at all may be as short as eight seconds.

31. K. E. Moyer and B. von Haller Gilmer, "The Concept of Attention Spans in Children," *Elementary School Journal* 54, no. 8 (April 1954): 464–66, state that young children may have an attention span of ten to fifteen minutes. More recent commentary suggests that attention spans for young adults may be as low at six to eight minutes. See, for example, Roderick G. W. Chu, Thoughts from New Albany (weblog), "Attention Spans and Learning," March 8, 2009 (http://rodchu.blogspot.com/2009/03-/attention-spans-learning.html).

32. A. H. Johnstone and F. Percival, "Attention Breaks in Lectures," *Education in Chemistry* 49, no. 13 (1976): 49–50.

33. Another way to explore the optimal length of a video is to look at how frequently videos are watched. To do so, I turned to YouTube and analyzed the duration of the videos found in the "most popular" category on April 20, 2008. Most of the videos were shorter than five minutes; the longest was eleven minutes. Disregarding the possibility that some videos might be more interesting than others, I calculated the average duration of the videos, weighted by the number of views. The weighted average duration of the most often watched YouTube videos is roughly four and a half minutes. Because some of the YouTube videos are much longer than the news stories found on the *New York Times* website, I viewed the videos that were longer that five minutes. It turned out that the long YouTube videos tended to be segments of music performances. The studies of *New York Times* and YouTube videos suggest that the most frequently watched videos, other than entertainment videos, are less that five minutes long.

34. The biases associated with jumping to a conclusion or solution are identified by numerous authors, for example, Aaron T. Beck and others, *Cognitive Therapy of Depression*, Guilford Clinical Psychology and Psychopathology Series (New York: Guilford Press, 1979), and Ian Mitroff's *Smart Thinking for Crazy Times: The Art of Solving the Right Problems* (San Francisco: Berrett-Koehler, 1998). For further information on cognitive dissonance, Wikipedia offers a useful summary.

INDEX

Poling, Harold, 54, 55, 56. *See also*
    Ford Motor Company
Porter, Michael, 25
Power, 140n22
Prospect theory, 138n8
Public Broadcasting System (PBS),
    130

Quality circles, 54, 55
Qumu, 106

Raps, A., 3–4
Respect, 52
Rhetoric (Aristotle), 48
Rumors, misinformation, and misun-
    derstandings, 34, 49, 51, 66

S&P *500*, 119, 120–21, 122f
Scripts, 83–84
Security issues, 105, 129
SharePoint (Microsoft), 106–07
Shakespeare, William, 50
Skype, 60, 61
Social services, 129–30
Software, 106–07, 108, 109
Sophocles, 50
Stocks and stock prices. *See* Financial
    performance
Stories and storytelling, 86, 123
Style, 81, 86–87, 88
Survey Monkey, 108
Surveys, 107–08
Symbolism, 81, 88

Technology. *See* Information technol-
    ogy
Telepresence, 61. *See also* Cisco
    TelePresence
Terminology and language, 86–87

Tracey, William: as CEO of MandA-
    Co, 2, 14, 19; challenge of adopt-
    ing ChangeCasting, 112–13;
    experience with ChangeCasting,
    116–18, 119, 125; launch of
    change effort, 35–36, 43–44. *See
    also* MandACo
Trust and trustworthiness: anony-
    mous feedback and, 66–67, 107,
    108, 133–34; being "real,"
    74–75; building trust, 47–53;
    compassion and, 85–86; protect-
    ing the conversation and, 65–66.
    See also ChangeCasting; Under-
    standing
Twitter, 60

UAW. *See* United Auto Workers
Uncertainty: as an amplifier of fear,
    32–33, 37; commitment approach
    and, 46; compliance approach
    and, 41, 44; listening and, 52;
    protecting the conversation and,
    51; trust and understanding and,
    47, 49
Understanding: change and, 47, 52;
    ChangeCasting and, 64–65,
    73–74, 77–78; communication
    and, 62; creation of, 73; symbol-
    ism and, 81. *See also* Change-
    Casting; Trust and trust-
    worthiness
United Auto Workers (UAW), 54, 55
United States. *See* Government agen-
    cies
Universities, 130

Video cameras and recorders, 96, 97,
    102, 103